Materials and Infrastructures 2

Research for Innovative Transports Set

coordinated by
Bernard Jacob

Volume 5B

Materials and Infrastructures 2

Edited by
Jean-Michel Torrenti
Francesca La Torre

iSTE

WILEY

First published 2016 in Great Britain and the United States by ISTE Ltd and John Wiley & Sons, Inc.

ISTE Ltd
27-37 St George's Road
London SW19 4EU
UK

www.iste.co.uk

John Wiley & Sons, Inc.
111 River Street
Hoboken, NJ 07030
USA

www.wiley.com

Library of Congress Control Number: 2016939898

British Library Cataloguing-in-Publication Data
A CIP record for this book is available from the British Library
ISBN 978-1-78630-031-7

Contents

Preface

The transport sector is very much concerned about environmental adaptation and mitigation issues. Most of these are related to the objective of curbing GHG emission by 20% by 2020, alternative energy and energy savings, sustainable mobility and infrastructures, safety and security, etc. These objectives require the implementation of advanced research works, to develop new policies, and to adjust education and industrial innovations.

The theme and slogan of the Transport Research Arena held in Paris (TRA2014) were respectively: "Transport Solutions: From Research to Deployment" and "Innovate Mobility, Mobilise Innovation". Top researchers and engineers, as well as private and public policy and decision–makers, were mobilized to identify and take the relevant steps to implement innovative solutions in transport. All surface modes were included, including walking and cycling, as well as cross modal aspects.

Policies, technologies and behaviors must be continually adapted to new constraints, such as climate change, the diminishing supply of fossil fuels, the economic crisis, the increased demand for mobility, safety and security, i.e. all the societal issues of the 21st Century. Transport infrastructures and materials, modal share, co-modality, urban planning, public transportation and mobility, safety and security, freight, logistics, ITS, energy and environment issues are the subject of extensive studies, research works and industrial innovations that are reported in this series of books.

This book is part of a set of six books called the *Research for Innovative Transports* set. This collection presents an update of the latest academic and applied research, case studies, best practices and user perspectives on transport carried out in Europe and worldwide. The presentations made during TRA2014 reflect on them. The TRAs are supported by the European Commission (DG-MOVE and DG-RTD), the Conference of European Road Directors (CEDR), and the modal European platforms, ERRAC (rail), ERTRAC (road), WATERBORNE, and ALICE (freight), and also by the European Construction Technology Platform (ECTP) and the European Transport Research Alliance (ETRA).

The volumes are made up of a selection of the best papers presented at TRA2014. All papers were peer reviewed before being accepted at the conference, and were then selected by the editors for the purpose of the present collection. Each volume contains complementary academic and applied inputs provided by highly qualified researchers, experts and professionals from all around the world.

Each volume of the series covers a strategic theme of TRA2014.

Volume 1, *Energy and Environment*, presents recent research works around the triptych "transports, energy and environment" that demonstrate that vehicle technologies and fuels can still improve, but it is necessary to prepare their implementation (electro-mobility), think about new services and involve enterprises. Mitigation strategies and policies are examined under different prospective scenarios, to develop and promote alternative fuels and technologies, multi-modality and services, and optimized transport chains whilst preserving climate and the environment. Evaluation and certification methodologies are key elements for assessing air pollution, noise and vibration from road, rail and maritime transports and their impacts on the environment. Different depollution technologies and mitigation strategies are also presented.

Volume 2, *Towards Innovative Freight and Logistics*, analyzes how to optimize freight movements and logistics, introduces new vehicle concepts, points out the governance and organization issues, and proposes an assessment framework.

Volumes 3 and 4 are complementary books covering the topic of traffic management and safety.

Volume 3, *Traffic Management*, starts with a survey of data collection processes and policies and then shows how traffic modeling and simulation may resolve major problems. Traffic management, monitoring and routing tools and experience are reported and the role of traffic information is highlighted. Impact assessments are presented.

Volume 4, *Traffic Safety*, describes the main road safety policies, accident analysis and modeling. Special focus is placed on the safety of vulnerable road users. The roles of infrastructure and ITS on safety are analyzed. Finally railway safety is focused upon.

Volume 5, *Materials and Infrastructures*, split into two sub-volumes, investigating geotechnical issues, and pavement materials' characterization, innovative materials, technologies and processes, and introducing new techniques and approaches for auscultation and monitoring. Solutions to increase the durability of infrastructures and to improve maintenance and repair are shown, for recycling as well as for ensuring the sustainability of the infrastructures. Specific railways and inland navigation issues are addressed. A focus is put on climate resilient roads.

Volume 6, *Urban Mobility and Public Transport*, highlights possible innovations in order to improve transports and the quality of life in urban areas. Buses and two-wheelers could be a viable alternative in cities if they are safe and reliable. New methodologies are needed to assess urban mobility through new survey protocols, a better knowledge of user behavior or taking into account the value of travel for public transport. The interactions between urban transport and land planning are a key issue. However, these interactions have to be better assessed in order to propose scenarios for new policies.

Bernard JACOB, Chair of the TRA2014 Programme Committee

Jean-Bernard KOVARIK, Chair of the TRA2014 Management Committee
March 2016

Acknowledgments

The European Commission, DG MOVE and RTD, the Conference of European Road Directors (CEDR), the European Road Transport Research Advisory Council (ERTRAC), the European Rail Research Advisory Council (ERRAC) and the European technology platform WATERBORNE-TP are acknowledged for their support and active contribution to the Programme Committee of the TRA2014, in charge of reviewing and selecting the papers presented at the conference, which forms the main input of this volume.

The French Institute of Science and Technology for Transport, Development and Networks (IFSTTAR) is acknowledged for having organized the TRA2014, in which 600 high-quality papers were presented, successfully.

Anne Beeldens, Pierre Marchal, Manuel Pereira, and Jon Krokeborg; coordinators of the topic on Materials and Infrastructure; all the other members of the Programme Committee; the reviewers who actively contributed to review and select the papers; and the authors who wrote them are acknowledged for their great job that produced the material for this volume.

Joëlle Labarrère, secretary of the Programme Committee of TRA2014, is acknowledged for her valuable help to the editors and for her support to prepare this volume.

Francesca La Torre

Professor Francesca La Torre is a Full Professor of roads, railways and airports at the University of Florence (Italy). She has been working in the field of transportation infrastructures for over 20 years. She obtained her PhD in 1998 at the University of Rome and she served as an assistant researcher at the University of

Illinois at Urbana-Champaign (USA). She is a member of the EC Horizon 2020 advisory group for “Smart, Green and Integrated Transport” and the infrastructures representative for academia in ERTRAC.

Jean-Michel Torrenti

Jean Michel Torrenti is the R&D director of the Materials and Structures Department of IFSTTAR. He is also professor at Ecole Nationale des Ponts et Chaussées. His research concerns mechanics of concrete and its coupling with durability aspects: behavior of concrete at early age, creep, leaching. It is applied to model the behavior of structures such as bridges, nuclear power plants and nuclear waste storage. He is the co-author of several books concerning concrete and concrete structures.

Bernard Jacob

Bernard Jacob, chair of the Programme Committee of TRA2014, is deputy scientific director for transport, infrastructures and safety with IFSTTAR. His research works are in bridge and road safety, traffic loads on bridges, heavy vehicles and weigh-in-motion. He has coordinated a number of European and International research projects. He is an active member in several scientific and technical committees (OECD/ITF, PIARC, TRB, etc.) and provided expertise to the European Commission. He is professor at Ecole Nationale des Travaux Publics de l’Etat and the president of the International Society for WIM (ISWIM). He has published more than 100 scientific papers and edited 10 published volumes of international projects and conference proceedings.

Introduction

The infrastructures of the future will have to be sustainable, seamless, resilient and durable, will respect the principles of circular economy and will have to be easy to monitor and manage. New technologies are currently available or under development to reduce the carbon footprint of infrastructures and to increase the overall sustainability and recyclability of transport while maintaining the utility and value of the infrastructures. However, the impact of these new solutions will only be effective once these are thoroughly disseminated and extensively deployed.

This volume presents a series of the most promising solutions and aims at disseminating them to improve the performances and efficiency of materials and infrastructures, through a choice of updated papers from the TRA2014 Conference. Selection is primarily based on a quality criterion, also taking into account the geographical diversity of papers in order to restore the originality and richness of current research.

I.1. Main findings

The papers contained in this volume demonstrate how technological solutions and new design and management methodologies can be implemented in different surface transport modes (roads, railways and waterways) to increase transport sustainability by improving infrastructures design, maintenance, recyclability and management. Both theoretical research and practical case studies explore topics such as characterization of pavements, bridges and soils, use of recycled and warm mix asphalts as well as high-performance materials to increase durability or to reduce the noise impact.

New management techniques for improving infrastructure resilience both roads and railways is a very timely topic that has been selected by the European Commission and the U.S. Department of Transportation as the subject of further Euro-American cooperation. This topic is extensively covered in this volume for a number of different transport modes.

Road infrastructures are typically "low technology" structures but timely, cost-effective and seamless monitoring is essential for the implementation of effective maintenance and management concepts. New solutions for pavement and soil characterization are being developed by implementing seamless technologies. These range from well-established techniques, such as ground penetrating radars (GPR) and weigh-in-motion (WIM) techniques, to innovative radar remote sensing techniques.

The development of new pavement materials is always a key topic for road and airport engineers and the implementation of recycled materials and warm mix asphalt will be the standard solution of the future. However, there is still a strong need for understanding the long-term performance of these materials *in situ* and for developing performance models that the designers can implement for adopting these technologies. This volume will help the designers and road managers interested in implementing these solutions and presents different case studies that will make the potential users feel more confident.

It is interesting to observe that infrastructure performances often conflict and therefore solutions such as porous asphalt, that can be very effective for noise reduction, is more sensitive to climatic changes due to the effect of freeze-thaw cycles.

Durability and maintenance are core issues for road researchers with the final aim in mind that the road of the future will have to be "Forever Open". However, local authorities are often faced with the issue of effective day to day maintenance. Infrastructure research too often focuses on highly trafficked motorways or primary road networks; therefore, it is extremely important that a research effort be specifically devoted to develop guidelines for the maintenance and repair of low volume roads, which represent a large portion of the whole road networks.

Railway and road infrastructures issues are usually tackled as separate but the recent work conducted by the joint roadmap for cross-modal transport infrastructure innovation toward a performing infrastructure has recently shown that a number of infrastructure research issues are cross-modal and therefore lessons can be learned across modes. This is clearly shown in this volume in which resilience to climatic changes covers both roads and railways and integrated modes are needed to achieve a truly resilient transport system.

This volume will be of interest not only for the research community and in higher education but also for professionals in the area of infrastructure design and management as well as economic and institutional decision makers. They will find state-of-the-art studies of key research issues, new advanced methods and illustrative case studies.

Volume 5 of the *Research for Innovative Transports* set is divided into two sub-volumes containing three parts each: five parts focus on roads but cover potentially cross-modal topics dealing with materials for infrastructures, auscultation and monitoring, durability and maintenance repair, recycling and sustainability issues and climate resilient roads. One part is specifically devoted to railways and inland navigation.

Sub-volume 1 contains parts 1–3. Part 1 deals with geotechnical issues and pavement materials' characterization. In this part researchers and practitioners can find new test methods and materials characterization techniques for non-conventional materials including recycled asphalt mixtures, warm mix asphalts but also fiber reinforced concrete materials.

Part 2 presents novel and high-tech solutions to monitor and assess pavement conditions to assist road authorities in this key management activity. These techniques include 3D mapping, remote sensing, GPR evaluation of pavement structural capacity and WIM monitoring solutions. The reader will also find a highly specialized study on integrating the electrical supply cables for public transport, for creating an electromagnetic induction field, in a prefabricated concrete slab.

Part 3 deals with the key road management issues of durability and maintenance repair. The recurrent theme of noise reduction has been tackled and designers and road authorities will be able to consider and compare the effectiveness of different solutions including non-conventional materials. Attention is also paid to noise issues in non-conventional analysis locations as level intersections in urban and rural areas. A very important issue for road managers is pothole repair. The guidelines developed in the POTHOLE project will be extremely helpful for local authorities looking for effective maintenance solutions.

Sub-volume 2 contains parts 4–6. Part 4 addresses recycling and sustainability issues, presenting case studies and full-scale tests. Asphalt recycling is a core issue for reducing the carbon footprint of transportation infrastructure. Road administrations and designers will find a very interesting overview of three transnational research projects on this topic as well as a case study from Slovenia.

Part 5 analyzes railways and inland navigation issues. New concepts for low maintenance and resilient infrastructure as well as optimizing operation and intermodal integration within the global transport system are proposed for technicians dealing with resilient infrastructure in any transport mode. Highly specialized railway experts will find studies on clip stiffness and on new innovative solutions for transition zones between the "normal" open tracks and "rigid" track sections. Waterways researchers will find an interesting new management approach to deal with suspended sediments.

Part 6 focuses on a key infrastructure issue of the future: resilience to extreme climatic conditions. Input from three continents (Australia, Europe and North America) highlight that this global issue needs trans-national solutions. An interesting overview of two transnational projects (RIMAROCC and SWAMP) introduces the topic followed by specific solutions adopted by single countries. The effect of climatic changes on pavements is assessed to answer questions of specialized pavement engineers.

I.2. Conclusions

This volume provides an insight on research, best practices and transport policies with a focus on state-of the-art advances in the fields of infrastructures and materials. The progress made in the implementation of new materials in pavement design as well as the evolution in the process of data collection and assessment, modeling and management, assisting academics, transport professionals, practitioners and decision makers to a better understanding of the current and future trends are demonstrated.

Future infrastructure monitoring techniques will be seamless, and this volume shows that there is a significant shift of the research world in this direction. These solutions now need to become current practices to really improve the transport system.

Reducing the infrastructure carbon footprint and increasing its resilience is possible but road managers and designers need to have design and management tools as well as case studies that will allow them to gain more confidence in the adoption of new and less impacting solutions.

PART 4

Recycling and Sustainability Issues

28

Introduction to European COREPASOL Project on Harmonizing Cold Recycling Pavement Techniques

Within the actual transnational road research program of the Conference of European Directors of Roads, the COREPASOL (Characterization of Advanced Cold-Recycled Bitumen Stabilized Pavement Solutions) project will be supported in the period 2013–2014. This project is realized by an international team from Czech Technical University in Prague, University of Kassel, University College Dublin, Laboratório Nacional de Engenharia Civil, I.P. (LNEC) and industrial partner Wirtgen GmbH. The project focuses mainly on harmonizing mix design of cold-recycled bitumen-stabilized materials following the existing scientific and engineering experience and approaches. The key objective in this relation is to develop and recommend comprehensive mix design and characterization by studying compaction methods, curing procedures and performance tests. Furthermore, the overall focus of the COREPASOL project's partial results is on the comparison of compaction methods used for the preparation of test specimens for cold-recycled mixes.

28.1. Introduction

Expected yearly worldwide demand for road rehabilitation works concerns about 1.7 million km of roads. This creates on the one hand a large need for natural resources used and on the other hand bears significant potential for the reuse of existing pavement materials and the focus on recycling techniques. Following the key characteristic of asphalt pavement being 100% recyclable, suitable techniques should be continuously developed and supported. Nevertheless, in the development on pavement structures and new technical solutions enhancing higher recyclability

---

Chapter written by Jan VALENTIN, Jan SUDA, Zuzana FORMANOVÁ, Konrad MOLLENHAUER, Michael ENGELS, Fátima BATISTA and Ciaran MCNALLY.

of existing structures, increased focus on environmental aspects, sustainability, durability and the protection of health and safety has been stressed during past decades. The use of recycled materials in this respect is a key strategy in developing sustainable road construction practices. Cold recycling techniques, especially if done in-place, can be seen as a highly progressive approach that addresses all of the stated targets for road construction – safe road surface conditions, high bearing capacity and durable pavements.

Cold recycling techniques are generally available and well known; nevertheless, they differ strongly in terms of utilization ratio, state of the art and public interest across Europe. Technical harmonization and European-wide guidance for best practice and effective performance design are needed. This includes guidance for designers as well. Maintaining the good infrastructure quality across Europe is one essential backbone to the performance of the European economy. Nowadays, there is no such harmonized mix design. Cold recycling, of course, is well known, but large differences in techniques and national states of the art were documented in the European-funded Direct-Mat project. Predominantly in-place recycling techniques are used with the application of solely bituminous binder or its combination with a hydraulic binder (cement and lime), especially for increased pavement bearing capacity. The mix design is usually based on defining optimum water content using the modified Proctor standard test and analyzing mechanical properties (compressive strength or indirect tensile strength (ITS)) after odd curing periods.

Cold recycling can be seen as a technology where several benefits and added values are linked – natural resources can be reduced, energy consumption decreased, road infrastructure protected from excessive construction-related transport and rehabilitation works can be shortened. Simultaneously, using this technology, the multiple recycling of already once recycled roads bears other potential and beneficial effects with similar advantages as described previously. The question is only the rate of old binder activation. If proper mix design is applied and suitable binders are used including activating the bituminous binder in reclaimed material, the new structure will lead to increased bearing capacity and improved pavement durability. Despite these facts, not more than 35% of reclaimed asphalt is reused by cold recycling.

28.2. Background of European project COREPASOL

Following the focus of the Conference of European Directors of Roads (CEDR) Transnational Road Research Programme 2012, COREPASOL (Characterization of Advanced Cold-Recycled Bitumen Stabilized Pavement Solutions) concentrates on research objectives defined within this programme for optimizing the recycling of pavements. Assessing European cold recycling asphalt pavement techniques and specific national designs and characterization should lead to suitable harmonization and introduction of a performance-based design concept applicable in most European countries. This should further increase the potential in using this set of techniques and allow an expert comparison of approaches done in different parts of Europe.

The COREPASOL project focuses mainly on harmonizing mix design of cold-recycled bitumen-stabilized materials following the existing scientific and engineering experience and approaches. The key target in this relation is to develop and recommend comprehensive mix design and characterization by studying compaction methods, curing procedures and performance tests, including the findings of previous European research projects such as SAMARIS, Superior Cold Recycling (SCORE) and Direct-Mat. The output should be applicable to all variants of cold-recycled mixtures containing bituminous binders or combinations with other binders or alternative fines. Separately, a specific alternative of using cold recycling just with hydraulic binder will be touched on as well. Aspects considerable for defining the durability of cold-recycled mixtures will be evaluated and specified as another important part of the project activity. The existing practice focuses only on water immersion and stiffness. Both characteristics are not a standard part in all mix designs known in Europe. Nevertheless to vindicate and promote cold recycling as an equivalent solution for hot mix asphalt, the durability must be predictable. For this reason, the project focuses on assessing stiffness, fatigue behavior, resistance to cracking and long-term moisture effects. Different test procedures will be evaluated and the most suitable approach is recommended including proposed threshold limits. This will make performance-based mix design possible.

Multiple recycling is addressed in several ways – recycling of existing pavement, where already recycled material is included, cold recycling as a base for repetitive rehabilitation with the same technique and applicability of other locally available granular materials. In this connection, the activity of reclaimed asphalt, impact of material aging and range of applicability not only in cold-recycled mixtures but also as a partial substitute of aggregates in hot mix asphalts will be specified and the most suitable practice recommended. The last perspective is closely related to technological solutions for the replacement of virgin non-renewable materials. Not only are applications such as aggregate substitutes touched upon but also the utilization of alternative binders (e.g. mechanically activated fly ashes or waste dust from aggregate production). Furthermore, using cold recycling in asphalt pavements leads to a reduction in binder application, whereas binder contained in reclaimed asphalt material is usually partially activated as well as influencing characteristics of the final mix. This phenomenon is applicable to second-generation recycling too. If hydraulic binders are used, they can be according to current experience and ongoing research works substituted, for example, by pozzolan or fly ash.

Furthermore, environmental aspects are addressed as well. Bitumen-based in-place cold recycling can be considered as the preferable solution to reuse materials containing tar by securing an immobilization effect. Similarly, it is expected that cold recycling will be an appropriate solution for asphalt mixes containing crumb rubber, which has been increasingly used in several countries for the last 5 years. In both cases, the risks related to reheating reclaimed material are eliminated.

28.2.1. *Objectives of the project*

The key objectives of the research are the definition and clear specifications of harmonized test, mix design and pavement design procedures for launching the discussion in CEN committees with the following points:

– harmonized advanced mix design procedures based on the traditional mechanical approach of analysis of performance properties for these types of mixes, applicable to all European countries;

– premises and possibilities for introducing performance-based mix design for more effective use of cold- recycled bitumen-stabilized materials;

– procedures to estimate material parameters feasible for the application in pavement design methods;

– determine the (multiple) recyclability of cold-recycled material with respect to the impact on mix design and characterization;

– suitable tools for the verification of environmental stability (leaching, tar detection and asbestos release) and for assessing the ecobalance of cold recycling techniques.

28.2.2. *Present situation on cold recycling*

As already stated before concerning cold recycling techniques, their characterization, related environmental issues and the material used represent an important task in several research activities, which have been followed during last 10–20 years worldwide. In Europe, several research projects realized within the 5th and 7th Framework Programs of European Union (EU) are focused on cold recycling characterization.

In the case of mix designs used for cold recycling mixtures, several research studies concerning cold recycling have emphasized that one factor that is hampering its widespread use is a lack of a suitable and harmonized design procedure. In the recent Direct-Mat project (7FP EU), it was concluded that cold in-place recycled mix design approaches and the applied test procedures still vary considerably among European countries [MOL 11]. Several issues have been identified as key points that need further investigation, especially:

– influence of inhomogeneities and different binders in the reclaimed asphalt pavement;

– effect of type and content of bituminous binder and corrective materials;

– evaluation of different mix design and mix characterization procedures, including laboratory compaction and accelerated curing test methods, key performance properties, quality control and assessment.

The SCORE project was a part of the 5th FP EU (from 2002 to 2005). Various aspects of recycling, from the characterization of milled materials to the formulation of recycled mixes and their properties, were studied. Stiffness and strength evolution of cold mixes made with emulsion and with foamed bitumen was evaluated under various curing conditions. The impact of the addition of a relatively small quantity of cement was also examined. Generally, in some countries, the more common application of cement with contents of 3–5% by mass has not been touched by the results. The research was focused on the *in situ* recycling of bituminous wearing courses; applications in base or binder courses have not been widely studied.

The assessment of performance-based characteristics commonly present for hot mix asphalt according to EN 12697-parts and for hydraulically bound and unbound materials according to EN 13286-parts represents a key part of the functional mix design but are in use for cold asphalt recycling materials only in limited extend and only in some countries. Generally mix design and mix quality is predominantly based on strength characteristics (comp. EN 13286-41, 13286-42). This is limiting the determination of cold recycling mix behavior for which both compressive and tensile properties are typical. For enhancing the mix design procedures, several countries have developed strategies separately from each other as follows:

– in France, the Duriez test is applied for the evaluation of voids content and compressive strength as well as stiffness (assessed at 15°C, resulting in stiffness values of about 1,500–4,000 MPa depending on the designed type I–IV) [ECK 08];

– in the UK, dynamic stiffness modulus has been assessed at 20°C resulting in values of about 1,500 MPa for the foam bitumen-stabilized materials (e.g. [KHW 01]);

– in Czech Republic, stiffness was assessed by IT-CY and 4PB test method at various temperatures resulting in values of 3,500–4,500 MPa (15°C) [VAL 09];

– in Germany, stiffness is evaluated by IDT with continuous loading according to EN 13286-43;

– in Portugal, usually requirements for cold bitumen emulsions mixtures refer to the unconfined compressive strength of cured specimens and their retained resistance after immersion in water [EST 12]. Furthermore, stiffness values in the order of 3,000 MPa at 20°C (IT-CY tests) have been reported in some research studies [BAT 04].

Besides the stiffness, additional performance-based properties are evaluated like shear strength properties as determined by triaxial test in South Africa [COL 08] or fatigue resistance. This was studied in the SCORE project by 2PB tests (problems with specimen preparation), in Czech Republic research studies by 4 PB fatigue test and indirect tension test fatigue tests [CHA 09], in Portugal research studies by IT-CY tests [EST 12] and by IDT in the UK ([KHW 01, FOR 02]).

With respect to environmental stability, it is important that the recycled materials have no negative impact on the environment. One possible way in which this can occur is through the leaching of contaminants through contact with water that infiltrates through the pavement. Leaching of hazardous substances and ecotoxicity are important issues to be assessed within the requirements of the new constructive products regulation. A review and some recommendations have been given in SAMARIS under the GROWTH program as part of the 5FP EU. Based on the work of TC292, leaching procedures for granular materials and sludge are described in standard-series EN12457. Applied procedures (column leaching test and one-stage leaching batch tests) are, nevertheless, not applicable to compacted mixes. No common threshold limits are prescribed.

28.3. Methods of cold-recycled asphalt specimen compaction

Requirements in the technical specifications of each country related to the compaction methods, specimen curing and the final evaluation of testing vary considerably, and there is no possibility of any simple comparison of the measured values. The specimens produced within the scope of this chapter were compacted with the most common compacting devices used worldwide – static pressure compactor, Marshall hammer and gyratory compactor.

The static pressure compaction is a relatively quick and simple method used in Germany, Czech Republic, Norway, Portugal and other European countries. According to the German procedure of the cold recycling mix compaction, the static axial pressure of 2.8 MPa is applied. The plastic behavior of the cold-recycled mixture leads to a stress depression, and therefore it is necessary to apply five to seven loading cycles until reaching the final maximum pressure of 2.6 MPa. Static compaction with double plunger is similarly used in Portugal. In this case, a stress of 21 MPa (170 kN) is applied for specimens of 101.6 mm diameter. Comparison tests on the same type of specimens have also been studied for 8 MPa loading. In Czech Republic, the specimens of cold-recycled mixtures are compacted in accordance with TP 208 by two pistons moving against each other. The applied pressure should be 5.0 MPa. During the compaction, it is necessary to repeatedly compensate the axial force until reaching the state when the power is stabilized for 30 sec at the value of 88.5 ± 0.5 kN. The specimens produced within the scope of this chapter were compacted according to the Czech methodology. According to the Norwegian methodology, one possibility is to use the static pressure of 4.5 MPa that is applied for 120 sec.

The second method of specimen compaction is to use a Marshall hammer that compacts the specimens by repeated loading with dynamic blows, usually with the same number gradually from both sides. This method is used, for example, in the United States, Australia and Poland; however, this compaction test method is on the wane. The specimens produced within the scope of this chapter were compacted by 75 blows of the Marshall hammer from each side, according to the requirement given in CSN EN 12697-30 for the specimens of 150 ± 1 mm diameter. Another widely used compacting device is the gyratory compactor. It is used for producing cylindrical specimens, which are used for the testing of some mechanical properties and determination of voids content or bulk density. The principle of the gyratory compaction is based on the combination of a slight static compression of the sample from the top and the simultaneously caused moving of the mold. The mold moves the way that its axis creates a conical rotating surface. The upper part of the mold is deflected by the top angle in the device, while the bottom part should be kept perpendicular to the axis of the created conical surface.

The French methodology determines for the cold-recycled mixtures the Duriez test. In this case, the specimens are compacted by the static pressure maintained for at least 5 min. There are two modes of the caused load (120 and 60 kN) according to the maximum grain size of the aggregates used. The loading rate is 1 $mm{\cdot}sec^{-1}$ and the maximum load must be obtained in the period between 5 and 60 sec. It is also necessary to allow the forced-out water to drain. A similar approach can be found in Ireland or the UK. On the other hand, Norwegian methodology prescribes compaction with angle of gyrations 1° at 600 kPa static pressure and standardized 30 gyrations/min. Usually, one set of specimens is prepared – test specimens compacted at number of gyrations that gives 96% of the density of 200 gyrations.

The use of the gyratory compactor for the asphalt specimen preparation occurred already in the 1930s. The first gyratory compaction device was used in Texas. After World War II, the use of a Marshall hammer was widely extended and the idea of using a gyratory compaction device was dropped. Nevertheless, the French scientists became interested in using a gyratory compactor already during the 1950. After a huge amount of research work in the 1960s and 1970s, the device was put into practice there. The use of the gyratory compactor in the United States was widely spread due to research program called "Strategic Highway Research Program", which took place in the years 1987–1993.

It should be noted that the qualities of the cold-recycled mixtures compacted by the gyratory compactor are affected by more factors than in the case of the hot mix asphalt compaction. First, these influencing factors occupy the water content. Water is drained away during the compaction, which prevents the accurate determination of the mix composition. Other very important factors are the type and the content of the binder used and also the aggregate parameters. It is also very difficult to find the correlation of the compactibility of the cold-recycled mixes produced in the laboratory and the mixes paved *in situ*. This is one of the frequent topics of current experimental studies and scientific papers as well.

Specimens prepared within the activities done by the Czech Technical University were compacted by Troxler Superpave Gyratory Compactor, Model 4140, according to CSN EN 12697-31. The inclination angle of the mold was set to 1.25° and the speed was 30 revolutions per minute. The specimens were compacted by the vertical pressure of 600 and 900 kPa, both loads gradually in combination with 40, 60, and 80 revolutions.

28.4. Experimental comparison of compaction methods

For the comparison of compaction methods, four cold-recycled mixes were designed, whereas in this chapter, the results of two are presented. The mix composition of these two mixes is presented in Table 28.1. Mix A contains cationic slow-breaking bituminous emulsion C60B7, which is commonly used in Czech Republic. Mix B consists of foamed bitumen, which was produced by the Wirtgen WLB10S device. Within the foamed bitumen, production water is injected into the 170°C hot bitumen (70/100) and that leads the foaming effect. The foamed bitumen is then immediately dosed into the two-spindle mixing device with controlled mixing speed, the Wirtgen WLM 30. Additional mixes C and D are similar to presented mixes and differ only in the absence of hydraulic binder.

The foamed bitumen is characterized by the expansion ratio (ER, mL/g) and the half-time of the foam settlement $\tau½$ (sec). Both parameters are strongly dependent on the kind and origin of the bituminous binder, the amount of compressed air added and the pressure of the water injected into the hot bitumen.

The intensity and efficiency of the foaming effect can be influenced by the basic physical conditions such as temperature, moisture and pressure. An optimal amount of the foaming water was specified in order to achieve the maximal expansion ratio (reached value ER = 18) and the maximum half-time foam settlement (reached value of 12 sec), according to the bitumen content. For the optimization of foaming water, the approach defined by Wirtgen manual was used.

	Mix A (%)	Mix B (%)
Reclaimed asphalt mix	91.0	88.5
Water	2.5	4.0
Bituminous emulsion	3.5	–
Foamed bitumen	–	4.5
Cement	3.0	3.0

Table 28.1. *Mix designs used*

In total, more than 100 cylindrical specimens were compacted from each mix by using the static pressure compactor, the Marshall hammer and the gyratory compactor. For some variants, repeated evaluation has been done. The diameter and height of the specimens were 150 ± 1 and 60 ± 5 mm, respectively. In order to know the degree of compaction, the bulk density from the sample dimensions and its weight were calculated for each sample. The main focus of the experimental testing was then based on the systematic measurement of the ITS and the stiffness modulus. Both values were measured on specimens, which were cured for 7 and 14 days in the air with temperature 20 ± 2°C and also on specimens cured for 7 days in the air and then immersed for 7 days in water.

28.4.1. *Bulk density*

Figures 28.1 and 28.2 show values of the ITS after 7 and 14 days. Furthermore, they also show the bulk density values of tested specimens, which had identical composition but were compacted by the different method. Bulk density is a very important parameter and the other measured characteristics are related to it. Therefore, the columns representing particular compacting methods are also sorted according to the bulk density value. As the figure shows, the bulk density value of the specimens produced by the gyratory compactor increases with the higher pressure and the number of revolutions. The bulk density values of the specimens compacted by the static compactor and Marshall hammer move somewhere in the middle of the achieved range. Bulk density of the specimens compacted by the dynamic loading of the Marshall hammer is lower than the bulk density of statically compacted specimens. In terms of the bulk density testing, it is quite interesting to compare the selective standard deviation values of the bulk density obtained from the dimension and weight of the tested specimens (Table 28.2). Despite the apparent material heterogeneity and the influence of water contained in the mix, the variations of the bulk density values are relatively accurate and comply with the measurement conformity.

According to actual technical specifications valid in the Czech Republic a preferred method for compacting test specimens is considered static pressure by a suitable hydraulic press. In the past, it was possible to also prepare specimens by the Marshall hammer. However, this method has not been used for a few years in Czech Republic. Introduction of optimal method of specimen compaction in the Czech technical specification, which would in best way simulate the compaction process during real site conditions, was forerun by many previous test. Despite this, it might be beneficial to supplement bulk density values of laboratory specimens by bulk densities of cores gained from construction sites and validate this argument. So far, however, this has not been done within the experimental activity discussed in this chapter. Such a comparison is planned as a next step within the key task of ongoing research.

From the presented results, it is possible to reach a similar bulk density of test specimens for the same mix design prepared by the different compaction method. This is valid for mix bonds by bituminous emulsion as well as by foamed bitumen. The most proximity values of bulk density for specimens compacted by static pressure with 5 MPa are reached on specimens compacted by 2 × 75 blows of a Marshall hammer or on specimens compacted by gyratory compactor at 600 kPa and 80 revolutions.

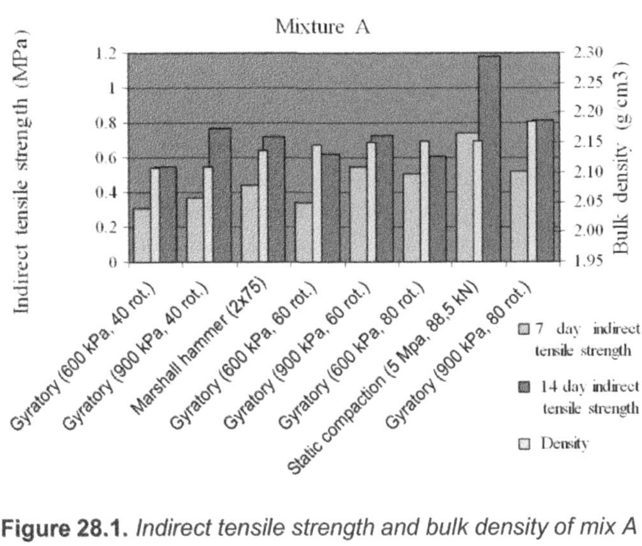

Figure 28.1. *Indirect tensile strength and bulk density of mix A*

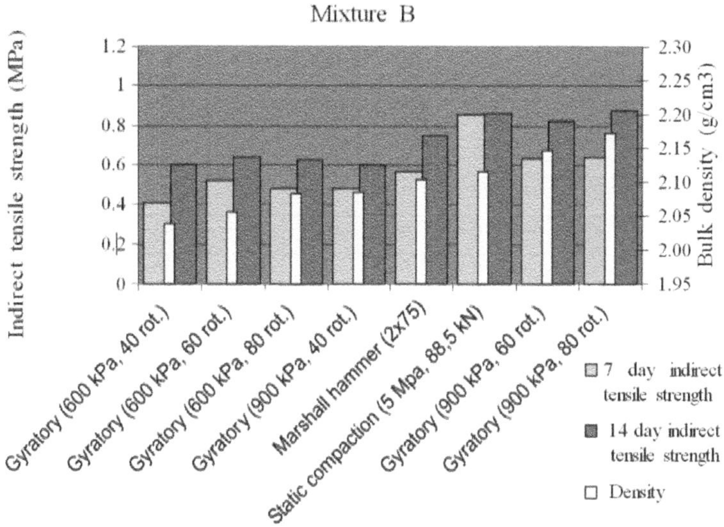

Figure 28.2. *Indirect tensile strength and bulk density of mix B*

28.4.2. *Indirect tensile strength*

Regarding the values of ITS, the situation is different if compared to findings for bulk density. This expected difference was the key motivation for carrying out extensive comparison testing. The highest ITS values are unambiguously always reached for static pressure compaction. These values are in some cases even more than 100% higher than ITS values for specimens with similar bulk density but prepared by the different compaction method. Therefore, it can be clearly stated that it is not possible to easily replace one compaction method with another compaction method.

In Table 28.2, the results of ITS as well as the ratios of water susceptibility are shown. All specimens fulfilled the requirements given by Czech technical specifications; nevertheless, again the influence of different compaction methods was evident. Respecting the fact that the specimens were done from the same mix (same mix design), these variations caused by the used compaction methods are relatively outstanding. For some specimens (e.g. specimens prepared by gyratory compactor with 900 kPa vertical pressure and 80 revolutions), quite a big increase in ITS has been found after curing in saturated state (water); for another set of specimens, the ITS decreased after curing in water (gyratory compactor with 600 kPa pressure and 80 revolutions).

Specimen compaction		Bulk density		Indirect tensile strength (MPa) after curing						Water susceptibility (-)	
				7 days dry		14 days dry		7 days dry + 7 days saturated			
		Mix A	Mix B	Mix A	Mix B	Mix A	Mix B	Mix A	Mix B	Mix A	Mix B
Marshall hammer		2.110	2.100	0.44	0.57	0.72	0.75	0.55	0.65	1.25	1.14
Gyratory compactor 600 kPa	40	2.108	2.040	0.31	0.41	0.55	0.60	0.40	0.46	1.29	1.12
	60	2.146	2.060	0.34	0.52	0.62	0.64	0.43	0.49	1.26	0.94
	80	2.152	2.080	0.51	0.48	0.61	0.63	0.45	0.44	0.88	0.92
Gyratory compactor 900 kPa	40	2.110	2.090	0.37	0.48	0.77	0.60	0.49	0.44	1.32	0.92
	60	2.151	2.150	0.55	0.63	0.73	0.83	0.61	0.62	1.11	0.98
	80	2.186	2.170	0.52	0.64	0.81	0.88	0.70	0.64	1.35	1.00
Static pressure		2.153	2.120	0.74	0.86	1.18	0.81	0.98	0.69	1.32	0.80

Table 28.2. *Indirect tensile strength and water susceptibility*

28.4.3. *Stiffness modulus*

Although the Czech technical specifications do not require validation of stiffness modulus for cold- recycled mixes, it represents an important characteristic with good potential for the future. This is given especially by the predicative level of this characteristic with respect to the description of the mix behavior. Simultaneously, stiffness represents an important parameter for pavement structural design. Therefore, it was decided to evaluate for all test specimens the stiffness modulus at 15°C by using the test method of repeated indirect tensile stress (IT-CY).

Gained values are summarized in Figures 28.3 and 28.4. The ranking of columns is similar to values for ITS values according to the bulk density value. An interesting finding is that compared to ITS values for stiffness, the difference between specimens with similar bulk density is not so striking, especially if observing the mixes bonded by bituminous emulsion (mix A). For example, basically similar stiffness modulus values for specimens with the same bulk density can be pointed out (specimens of mix A compacted by static pressure and specimens compacted using gyratory compactor with 60 revolutions and static load of 900 kPa). Also in the case of stiffness modules for mixes with foamed bitumen (mix B), the differences in stiffness for mixes compacted by static pressure and for mixes compacted by other methods are not as distinct as is evident for results of ITS.

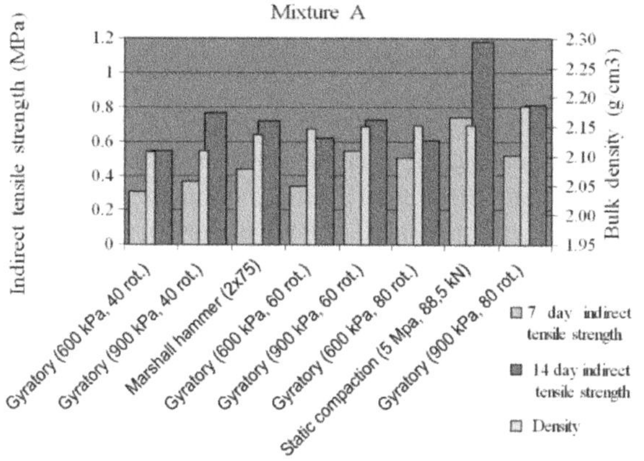

Figure 28.3. *Stiffness modulus and bulk density of mix A*

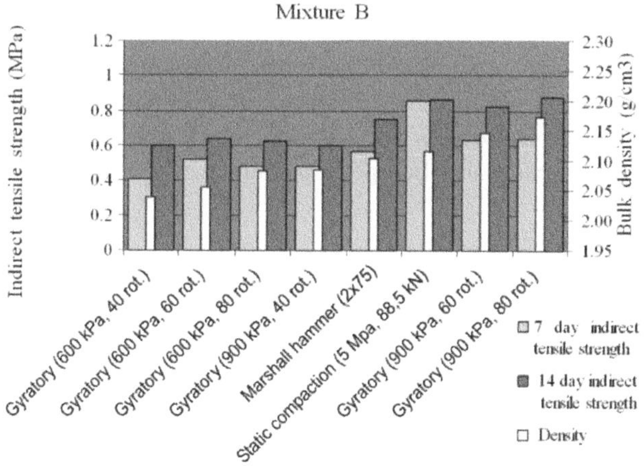

Figure 28.4. *Stiffness modulus and bulk density of mix B*

28.5. Publication policy

Presently, there are in Europe several ongoing or already realized international R&D projects focusing on the analysis and specification of differences in approach different European countries are using if evaluating the quality of asphalt cold-recycled mixes or similar mixes with different types of input recyclable materials. The analyses mainly look at criteria on input materials, methods used for preparing test specimens and sets of empirical and performance-based characteristics used. In the case of cold recycling techniques, it is expected for the future to set optimal methodology, which would be possible to implement in all European countries.

With respect to the requirements of national technical specifications, in particular countries concerning the preparation of test specimens for cold-recycled mixes the approaches are different. Usually, compaction by static pressure or gyratory compactor is used. The use of a Marshall hammer in Europe for this type of mixes is currently in decline. Differences in specimen compaction, nevertheless, not only concern the apparatus used but are also related to applied loading, number of revolutions, number of applied blows, etc. The method for the preparation of test specimens has a crucial impact on bulk density and other characteristics currently assessed (indirect tensile strength and stiffness). From a desire to record all differences cased by the compaction method used, a comparative study has been carried out with results summarized in this chapter. Within this study, it was confirmed that there will be an evident benefit of introducing a unified methodology instead of continuing in the present state, which is marked by incomparability of values received in different countries by different methods used for specimen compaction. In the near future, it is expected that this research will be extended by comparing values on laboratory-prepared specimens and cores gained from real construction sites in selected EU countries. Second, a similarly important task for further experimental activity is the assessment of different methodologies and conditions used by particular European countries in the field of test specimens curing.

28.6. Acknowledgments

This study has been supported by the research project COREPASOL carried out as part of the CEDR Transnational Road research Programme Call 2012. The funding for the research was provided by the national road administrations of Belgium (Flanders), Denmark, Finland, Germany, Ireland, the Netherlands, Norway, Sweden and UK. This study has been further supported by the CTU student research project SGS13/125/OHK1/2T/11.

28.7. Bibliography

[BAT 04] BATISTA F., Novas técnicas de reabilitação de pavimentos – Misturas betuminosas densas a frio (Innovative pavement rehabilitation techniques – Dense asphalt cold mixtures), PhD Dissertation (prepared under the cooperation between LNEC and FEUP), Porto, Portugal, 2004.

[BOC 05] BOCCI M. *et al.* "Performance analysis of mixtures produced using foamed bitumen", *3rd Internartional SIIV Congress*, Bari, 2005.

[CHA 09] CHAMOT S., ROMERO P., "Fracture energy evaluation of cold in-place recycling mixtures", *Advanced Testing and Characterization of Bituminous Materials*, CRC Press/Balkema, Leiden, pp. 1123–1130, 2009.

[COL 08] COLLINGS D.C., JENKINS K.J., "Characteristics of materials stabilized with foamed bitumen", *Proceedings of 4th Eurobitume & Euroasphalt Congress, Copenhagen*, pp. 402–113, 2008.

[ECK 08] ECKMANN B. *et al.*, "Stiffness of cold recycled materials", *Proceedings of the 4th Eurobitume & Euroasphalt Congress*, Copenhagen, pp. 403–408, 2008.

[ECK 08] ECKMANN B., SOLIMAN S., "Performance assessment of cold recycling in place", *Proceedings of International Symposium on Asphalt Emulsion Technology*, 2008.

[EST 12] ESTRADAS DE PORTUGAL S.A., Caderno de Encargos Tipo Obra (Type Specifications for Construction), Estradas de Portugal (EP), S.A, 2012.

[FOR 02] FORDYCE D., KHWEIR K., "The use of a designed foamed bitumen stabilized RAP in an urban high street", *9th International Conference on Asphalt Pavements*, Copenhagen, 2002.

[HAR 01] HARMAN T., BUKOWSKI J., MOUTIER F. *et al.*, The history and future challenges of gyratory compaction 1939–2001, Report, Federal Highway Administration, 2001.

[HOS 09] HOSIN D.L., YONGJOO T.K., BYUNGHEE T.H., Laboratory performance evaluation of CIR-emulsion and its comparison against CIR-foam test results from phase II, Research report, University of Iowa, 2009.

[IWA 13] IWAŃSKI M., CHOMICZ-KOWALSKA A., "Laboratory study on mechanical parameters of foamed bitumen mixtures in the cold recycling technology", *Procedia Engineering*, vol. 57, pp. 433–442, 2013.

[KHW 01] KHWEIR K. *et al.*, "Aspects influencing the performance of foamed bitumen stabilized aggregate mixtures", *The Asphalt Yearbook*, Stanwell, pp. 27–34, 2001.

[VAL 09] VALENTIN J., Problems of selected performance characteristics of cold recycling mixes, PhD Thesis, CTU, Prague, 2009.

[VAL 10] VALENTIN J., MONDSCHEIN P., MACKO M. *et al.*, "Assessment of fatigue characteristics of cold recycling mixes by different test methods", *Proceedings of the 11th International Conference on Asphalt Pavements*, International Society for Asphalt Pavements, Lino Lakes, 2010.

[WIR 10] WIRTGEN GROUP, Cold Recycling – Wirtgen Cold Recycling Technology, Manual, 3rd ed., 2010.

29

Technical Performance and Benefits of Recycling of Reclaimed Asphalt Containing Polymer-modified Binder in Premium Surface Layers

The technical possibilities and benefits when recycling polymer modified asphalt (PMA) from surface layers into new high-quality surface layers using hot mix recycling are investigated. First rheological properties of blended binders are determined, consisting of virgin binders blended with extracted binders from reclaimed asphalt-containing polymer modification. The performance of asphalt mixtures containing reclaimed asphalt (RA) with polymer modified binder is tested with respect to water resistance, stiffness, rutting and fatigue. The economic and ecological effects are evaluated using lifecycle analyses (LCA) and lifecycle cost (LCC). It has been demonstrated that the polymer present in the RA still positively affects the material behavior of asphalt mixes produced with this material. LCA and LCC analyses clearly show the environmental and economic benefits of recycling of PMA. Based on these positive results, validation in practice is the next logical step.

29.1. Introduction

Throughout Europe, polymer modified asphalt (PMA) has been used extensively in the past decades for high trafficked roads and premium pavements, especially for surface layers. The addition of polymers contributes to the durability and functionality of these premium pavements, such as noise reducing pavements (the Netherlands) or rutting resisting pavements (Denmark and Slovakia). The first generation of these pavements is increasingly approaching their end of life.

Chapter written by Greet LEEGWATER, Jozef KOMAČKA, Gang LIU, Erik NIELSEN and Eva REMIŠOVA.

Therefore, the road sector is facing a rapidly increasing source of reclaimed asphalt (RA) that contains polymer modified bitumen (PMB). Depending on the aged state of the modified binder and the deterioration of the aggregate gradation during milling and further pre-processing, it is the challenge to the road sector to ensure – as far as possible – that the RA-containing PMB is recycled at its highest practical potential. The goal in recycling is not to achieve the highest possible recycling percentage, but to avoid downgrading RA containing a potential valuable asset such as PMB.

This chapter describes the work done in the RECYPMA project that was a part of the ERA-NET ROAD II program. The aim of this project was to investigate the possibilities for recycling PMA from surface layers into new high-quality surface layers using hot mix recycling. This chapter will give an overview of the results; more details can be found in the project reports [KOM 13, LEE 13, LIU 13a, NIE 13a, NIE 13b].

The chapter is organized as follows. First, the state of the art on recycling of RA containing polymer modified binder is described in brief. Next, the results of the laboratory tests are presented. For the tests, three different types of RA are used. The RA all originates from surface layers and contains styrene–butadiene–styrene (SBS) modification. Tests include the characterization of the extracted and blended binders using different test methods such as a dynamic shear rheometer (DSR) and infrared analysis. Based on the performance of the blend binders, mix designs are made for premium pavements that contain 15% and 40% RA containing polymer modification. The asphalt mixes are tested under laboratory conditions to establish their performance with respect to water resistance, stiffness, rutting and fatigue. Based on the established technical performance, the economic and ecological effects are evaluated using a lifecycle analysis (LCA) and a lifecycle cost (LCC) analysis. The chapter concludes with a summary of the obtained knowledge and an outline of work that is still needed before large-scale application of premium polymer modified surface layers containing RA can be realized.

29.2. State of the art on recycling of polymer modified asphalt (PMA)

29.2.1. *Focus and aim of literature review*

As the project focuses on recycling of surface layers into new high-quality surface layers, the state of the art portrays documentation that especially highlights the added value of polymers in RA. For basic hot mix recycling technology, readers are referred to the literature (e.g. [VAN 09]). The literature review has also gathered information on laboratory extraction methods, which is required for characterization of properties of an aged PMB. The following statements highlight the overall outline of the findings; more details can be found in [NIE 13a].

29.2.2. *Extraction and recovery*

The SBS-type polymer modification studied in this research is the most commonly applied polymer modification in hot mix asphalt. Based on a literature review and the response of leading European laboratories and research institutions to a dedicated questionnaire, the following two conclusions were drawn on extraction of polymer modified binder:

– for practical purposes, the European standards EN 12697-1 and 12697-3 can be used to recover a representative bituminous binder irrespective of the solvent used;

– in this project, dichloromethane (methylene chloride) is used as solvent for the extraction and recovery of SBS polymer containing bituminous binder from the RAs.

29.2.3. *Laboratory mixing*

A laboratory mixing procedure is formulated aimed at producing a representative asphalt mix containing recycled PMA. The method does not mimic full-scale production, since there is a huge difference in mixing efficiency and conditions in the asphalt plant compared to the laboratory mixer. The mixing procedure mentions the order of adding the different components and the considerations for determining the mixing temperature (depending on the properties of both the reclaimed and the virgin binder).

29.2.4. *Experience with utilization of RA-containing PMB in full-scale production*

As the use of PMB in hot mix asphalt on a major scale only started a few decades ago, it was only recently that a level of RA has been reached where selective recycling has become technically and economic feasible. As a result, the amount of well-documented experience found in the literature is limited. Due to this lack of knowledge, the literature review has resulted in a list of challenges that are recognized when aiming for recycling of PMA into surface layers:

– to be able to consciously “harvest” high-quality RA, information about the material is needed before milling. However, often existing documentation on road sections is poor, and as a result limited information is present on applied materials, such as binder and aggregate types;

– the materials have also degraded over time, binders age over time and aggregated sizes are expected to be reduced trough wear and milling. Both a very aged binder and a fine aggregate gradation may limit the recycling possibilities;

– in order to ensure proper mixing of the RA containing polymer modification with fresh materials, the RA has to be heated to high temperatures, resulting into the risk of burning materials;

– recycling above 15% is not feasible by cold mix addition of RA. As a result, high recycling levels require technological advanced asphalt plants, e.g. the presence of a double drum mixer;

– RA containing polymer-modified material is even more sticky compared to normal RA. This might lead to problems with handling and processing of the material in the asphalt plant.

29.3. Materials

29.3.1. *RA containing polymer modification*

Three "old" SBS PMB-containing asphalt mixtures were reclaimed from typical surface layers of premium pavements in three different countries: stone mastic asphalt (SMA11) in Denmark, porous asphalt (PA4/8) in the Netherlands and dense asphalt (AC11) in Slovakia. The SMA mix was 22 years old, the binder used was Caribit Plus 85, a SBS polymer modified bituminous binder with a penetration range 70–100 × 0.1 mm and a softening point (SP) above 75°C. The PA mix was 7 years old, and the binder used was a Styrelf PMB 40/100-65 HD. The AC was 15 years old, the binder used was Apollobit MCA-S, a SBS polymer modified bituminous binder with a penetration ranging from 50 to 100 × 0.1 mm and a SP above 70°C.

29.3.2. *Binder extraction and virgin binders*

Following European norms EN 12697-1 and EN 12697-3, extraction and recovery were performed on these three reclaimed mixtures to obtain three PMB-containing binders with abbreviations RAD (SMA from Denmark), RAN (PA from the Netherlands) and RAS (AC from Slovakia).

In order to evaluate the benefits of the high-quality RA, two types of virgin binders were used, a normal paving grade binder (PGB) and a polymer modified binder. The PGB is used to assess if the remainder of the polymer modification in the RA provides benefits for a new mix based on a PGB. The polymer modified binder is used to assess the influence of the presence of old polymer modified binder to the new mix. Virgin binders were selected for the new asphalt mixes assuming a recycling percentage of 15% and 40% using the "LogPen rule". This rule indicates the expected binder penetration of a mix of old and new binder. For the Dutch and the Slovakian mix, the following binders were used:

– *PGB1*: Q8, straight run bitumen, paving grade 70/100, provided by Kuwait Petroleum (Nederland) BV;

– *PMB1*: 70/100-83 (KR), modified bitumen by mixing 10% of D0243 SBS in B160/220 bitumen, produced by Kraton Polymers Nederland BV.

For the Danish mix, the following fresh binders are used:

– *PGB2*: 70/100, straight run bitumen produced by blending two different bitumen's (40/60) and (330/430) in the proportion of 71.2% and 28.8%, respectively;

– *PMB2*: 90/150-75, SBS PMB, provided by Colas Danmark A/S as a reference sample from their production of PMBs.

29.3.3. *Asphalt mixes*

The three different types of asphalts for wearing courses (SMA, AC and PA) were designed and produced for laboratory testing. Five combinations of virgin material and RA were used for each type of asphalt, based on three RA contents (0, 15, and 40%). An overview of the mixtures is presented in Table 29.1. The mixture with 0% RA is used as a benchmark. It was chosen not to test a mixture with PGB + 15% RA, irrespective of the polymer level in the RA; the total polymer level is too low to expect any effect of the polymer.

RA content	SMA 11		PA 8		AC 11	
	PGB2	PMB2	PGB1	PMB1	PGB1	PMB1
0 %	×	×	×	×	×	×
15 %	–	×	–	×	–	×
40 %	×	×	×	×	×	×

Table 29.1. *Matrix of tested asphalt mixtures*

Aggregate and fillers commonly used for the SMA11 production in Denmark, PA8 production in the Netherlands and the AC11 production in Slovakia were used in the mix design. More details on materials, mixture composition and mixture production can be found in the specific project report [KOM 13]. Mixing temperatures were based on viscosity measurements that were conducted on the binders [LIU 13a].

The binder content in RAs was determined by extraction procedure according to EN 12697-1. The obtained binder contents in RA are given in the following:

– RAN 5.42%;

– RAS 5.05%;

– RAD 4.60%.

29.4. Properties of extracted and blended binders

High shear mixing was used to blend the binders, and the aim was to obtain an optimal blend. Optimal blending will give an upper limit of the properties of the combined binder in mixtures, as it is expected that a lower level of blending will be achieved during mixing in practice.

Table 29.2 gives an overview of the binders that were studied in this research. Noteworthy is that the original RAD binder has also been characterized, as it was still available from storage. The characteristics of the binders are determined using a variety of rheological and chemical tests. The testing program on the recovered and blended binders included:

– penetration, SP;

– DSR master curves;

– viscosity measurements;

– Fourier transform infrared (FTIR) spectroscopy, showing chemical composition;

– gel permeation chromatography (GPC), providing particle size distribution.

In [LIU 13a], detailed information is provided on test methods and the complete test results. In this chapter, a short summary of the results will be presented; in [LIU 13b], a more extensive summary can be found.

From the characteristics of the binders extracted from the RA, it can be observed that, even though the SBS polymer degrades slightly over time, the influence of the polymer on the phase angle is still clearly present. Figure 29.1 clearly demonstrates this for the extracted RAD binders. In Figure 29.1, it can also be seen that the influence of the polymer remains present if the extracted binder is blended with virgin PGB bitumen or virgin PMB. Therefore, it is concluded that it is possible to restore the rheological properties of the reclaimed PMB binder to its original state by mixing it with a soft virgin PMB binder.

Binder		RA content (%)	Fresh bitumen content (%)
Reclaimed	RAD	100	0
	RAN	100	0
	RAS	100	0
Virgin	PGB1	0	100
	PGB2	0	100
	PMB1	0	100
	PMB2	0	100
	Original binder RAD	0	100
Blended RAD	15%RAD+85%PGB2	13.5	86.5
	40%RAD+60%PGB2	36.5	63.5
	15%RAD+85%PMB2	13.7	86.3
	40%RAD+85%PMB2	36.9	63.1
Blended RAN	15%RAN+85%PGB1	15	85
	40%RAN+60%PGB1	40	60
	15%RAN+85%PMB1	15	85
	40%RAN+60%PMB1	40	60
Blended RAS	15%RAS+85%PGB1	15	85
	40%RAS+60%PGB1	40	60
	15%RAS+85%PMB1	15	85
	40%RAS+60%PMB1	40	60

Table 29.2. *Overview of extracted, virgin and blended binders studied*

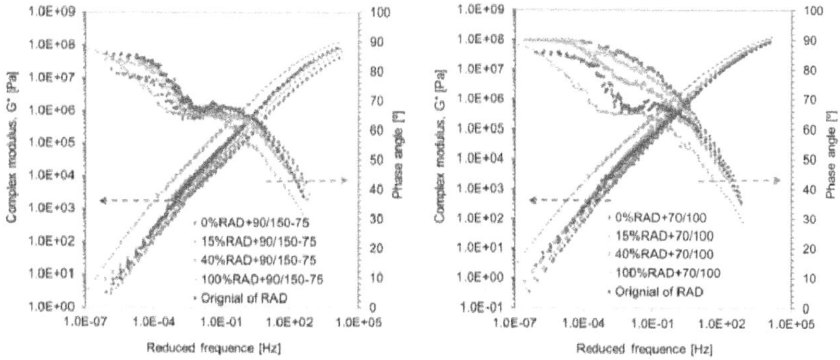

Figure 29.1. *Master curves of virgin, extracted and blended bitumen from Denmark (RAD): (left) when blending with PGB; (right) when blending with polymer modified bitumen. For a color version of this figure, see www.iste.co.uk/jacob/materials.zip*

In the mix design, a combination of the logPen model and the SP model is used, as they approximately predict the penetration value at 25ºC and the SP of the blended binder both for PGB and PMB. For a more detailed mix design, the Grunberg–Nissan model can be used. This model provides a more detailed description of the rheological properties of the binder, such as viscosity for a specific temperature and the complex modulus for a wide frequency range.

Due to the polymer modification and aging of the binder, the required mixing temperatures for the blended binders are high, in some cases above 180°C. As such high temperatures should be avoided to prevent degradation of the polymer, practical solutions to deal with this high mixing temperature requirement needs attention in future research.

The SBS index obtained with FTIR can be used to assess the relative SBS content of the extracted binder, which is useful to determine the level of modification required for the virgin modified binder. However, this method needs to be calibrated to match the type of SBS used in order to be accurate. GPC analyses can be used to the assess the degradation of the polymer, however this is only possible if the original binder is available as a reference.

29.5. Performance of asphalt mixtures with RA containing polymer modified binder

The aim of laboratory tests on mixtures was to investigate the impact of RA containing polymer modification on the properties of asphalt. The following tests were performed in order to assess the performance of the asphalt:

– water sensitivity test: the indirect tension strength ratio (ITSR) values show the ratio of material strength with and without exposure to water;

– wheel tracking test: the output of tests provides an estimate of the resistance to rutting;

– stiffness and fatigue test: the results provide an estimate for the structural life expectancy and will also give an impression on the integrity of the material.

All mentioned tests were carried out according to the relevant European standards. An overview of the results is presented in Table 29.3. Komačka *et al.* [KOM 13] provide detailed information on test methods and the complete test results. This chapter gives a short summary of the results; a more extensive summary can be found in [KOM 14].

Mixture	Virgin binder	RA content (%)	ITSdry	ITSwet	ITSR	Rutting resistance	Stiffness	Fatigue
SMA11	PGB	40	+	+	+	+	+	×
	PMB	15	+	+	-	=	+	
		40	+	+	-	=	+	
AC11	PGB	40	+	+	+	+	+	+
	PMB	15	+	+	-	=	+	–
		40	+	+	-	=	+	–
PA8	PGB	40	+	+	+	+	+*	×
	PMB	15	+	+	+	=	+	
		40	+	+	+	=	+	

+,better than benchmark mixture (without RA).

-, worse than benchmark mixture (without RA).

=, comparable to benchmark mixture (without RA).

x, not tested.

*Only for temperatures above 0 °C.

Table 29.3. *Summary of RA addition influence*

The ITSR values for all mixtures are high, which means that the water sensitivity of the mixtures containing RA meets the requirements for surface layers. The indirect tensile strength of the mixtures correlates well with the penetration values of the blended binder determined in the binder tests; this is shown in Figure 29.2. As optimal blending of the binder was targeted in the binder characterization, the good correlation indicates good blending of the binder in the asphalt mixture. Due to the extremely porous nature of the PA8 mixture, the correlation is not present for this material.

The wheel tracking test showed good performance with respect to rutting for all mixes containing RA. The mixture with PGB without RA showed a significant amount of rutting. The high rutting resistance of mixtures with RA might indicate an advantage of the remaining SBS in the binder; however, it could also be the effect of aging of the binder.

The mixtures with RA containing PMB have a higher stiffness compared to the mixtures without RA. This could be an advantage because a larger stiffness leads to a higher bearing capacity. However, if the stiffness is very high, this could also result in brittleness at low temperatures. Therefore, it is necessary to find a balance between necessary stiffness and a risk of brittleness taking into account temperature conditions at the intended locality. To further investigate this, a low temperature cracking and properties test according to EN 12697-46 should be performed in the future.

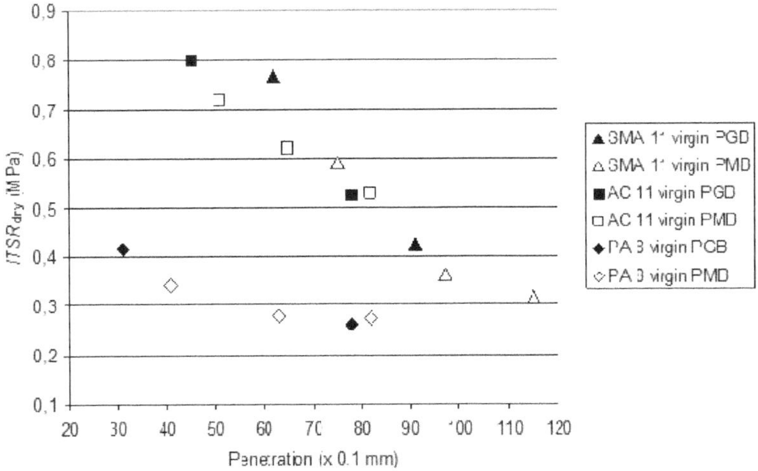

Figure 29.2. *Relation between penetration and indirect tensile strength (ITS) value, and AC11 and SMA 11 show a good correlation*

Fatigue test was only performed on the AC 11 mixtures. The mixture with the PGB and 40% RA showed better fatigue parameters compared to the mixture with paving grade bitumen and no RA. On the contrary, the mixture with virgin PMB was the most fatigue resistant and an addition of RA decreased fatigue performance and fatigue life. It is postulated that the SBS content in blended bitumen is an important parameter for fatigue performance. The remaining SBS in RAS binder increased fatigue resistance of the mixture with PGB. As the SBS content in RAS binder is lower (± 5%) compared to the SBS content of the virgin PMB (10%), the RA mixtures have a lower SBS content and therefore show an inferior fatigue performance. Further research is needed to validate this hypothesis.

Based on the test results presented in this chapter, it is concluded that recycling of RA from surface layers into new surface layers is expected to result in comparable asphalt performance for surface layers. However, this assumption still needs to be validated through more extensive research as a limited number of asphalt mixes is tested with respect to a limited amount of performance characteristics.

29.6. Microscopy investigations of RA and asphalt mixes containing RA

Microscopy analysis of thin and plane sections of asphalt has been performed as part of the research to serve two goals: to analyze the three RA materials used and to provide a visual assessment of the new asphalt mixes (with and without RA, with PGB and PMB) all microcopy images and observations can be found in the relevant project report [NIE 13b]. In this chapter, a short summary of the results is presented.

The plane sections give an impression of the buildup of the compacted asphalt mixture. The PA8 and AC11 mixtures are homogeneous, and the SMA mixtures show some inhomogeneities. As Marshall's compaction is used in this research, which is known to result in inhomogeneities for SMAs, this is not further investigated.

The thin sections show qualitatively the presence and extent of the dispersion of the SBS polymer from the typical yellow spots. As these spots can be observed in all RA materials, it is concluded that polymer is still present in the binder. The characterization of the extracted binder demonstrates that the polymer present also affects the binder response.

Thin sections made up of the asphalt mixes show some inhomogeneities in mixing. Parts of the mortar are homogeneous; however, as shown in the micrograph in Figure 29.3 at some locations, there is a clear distinction between old and new mortar. These inhomogeneities do not seem to affect the material behavior, as tensile tests on the asphalt showed a good correlation with the penetration of the binder (Figure 29.2). As a more homogeneous mixture is expected to lead to a better performance, future research using the microscope is recommended in order to investigate the homogeneity of full-scale plant-produced mixes to assess whether or not prolonged mixing times has the potential to improve the asphalt performance.

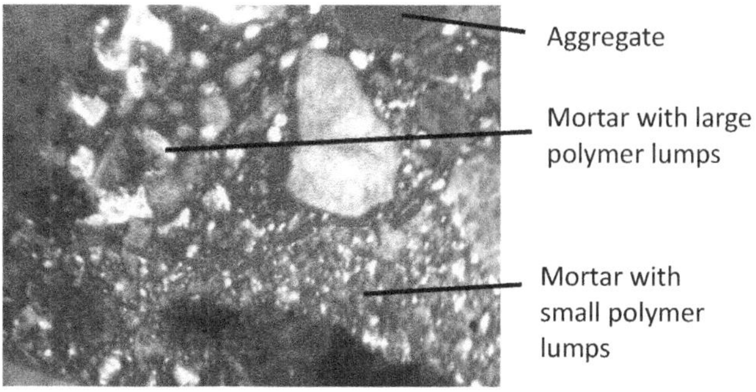

Figure 29.3. *Detailed image from a micrograph of AC 11 with PMB and 40% RA*

29.7. Environmental and economic benefits

By replacing virgin material, recycling clearly reduces the amount of resources needed and waste that is produced for new roads or road maintenance, as a result recycling reduces the environmental impact. Besides this, recycling also provides a cost reduction due to the fact that RA (including the needed pre-processing that is needed for recycling) is generally cheaper than virgin material. Polymer modifications are especially expensive and if the originally present polymers could also be beneficial in the recycled product, this would provide an interesting economical argument for recycling of polymer modified binders.

In order to quantify these potential benefits of recycling of PMA, an analysis of the environmental and economic impact of recycling is performed. In this analysis, a road with polymer modified binder is taken as a reference and compared to a road with recycled material with polymer modification. The basis for comparison is a 3.5 m wide pavement, representing a driving lane of a main road, over a length of 1 km during 50 years. The materials and production processes are based on the binders and mixes previously describes. Questionnaires were sent out to contractors and material suppliers to obtain cost and hauling distances. The life expectancy of the road, which is also a part of the LCA, is based on the performance of asphalt specimens in the laboratory.

In order to obtain a complete image of the environmental benefits of recycling of PMA, an LCA analysis has been performed. The total recycling process is considered, including all materials and processes. The economic benefits are assessed using a cycle cost analysis that includes all cost during the whole life span. The benefits of the different scenarios are presented in Table 29.4. The analyses show that the effect of using recycled material on the environmental impact and LCC is significant, especially if 40% recycling is realized. With respect to costs, it should be noted that the cost of the binder is responsible for up to 70% of the materials costs. The actual price level of binders can vary substantially over the years under the influence of the oil price.

Mixture	Virgin binder	RA content (%)	LCC costs (%)	LCA ReCIPe (%)
SMA 11	PMB	15	–7	–8
		40	–19	–22
AC 11	PMB	15	–5	–6
		40	–12	–21
PA 8	PMB	15	–5	–8
		40	–13	–23

Table 29.4. *Economic and ecological benefits of recycling of RA containing PMB into new polymer modifies asphalt mixes*

When the LCA method ReCIPe is used, the binder determines the environmental impact to a large extent: 40–55%, dependent on the exact scenario. The second most important parameter is transport. There are also two other methods available in Europe used to assess the environmental impact: EDIP 2003 and ecological scarcity. These methods attribute less impact to the binder and more to transport and the production of other raw materials; however, the resulting reductions that can be realized by recycling are similar compared to the results found with ReCIPe.

The assessed environmental impact is not very sensitive to variations in key starting points. However, service life assumptions have the large influence on calculated environmental impact. Therefore, quantification of the influence of RA on the service life of surface layers is very important and should be studied further.

29.8. Conclusions and recommendations

Binder research indicates that the extracted binder from the RA with PMB still contains an active part of polymer that influences the material behavior as can be expected from a polymer modification. Research on mixtures containing RA with PMB shows that, if proper mix design is used, a similar performance in surface layers is realized with respect to water sensitivity, stiffness and rutting resistance.

From an economic and ecological perspective, the benefits for National Road Authorities as well as the road sector are considerable because of the shorter hauling distances of aggregates and the reduction of primary resources needed. This is especially important for EU countries that are dependent on import of primary raw materials.

The research indicates that there are technical possibilities for recycling of polymer-modified surface layers into new surface layers and that the benefits are expected to be significant. However, still many challenges lie ahead before large-scale application is possible. Risks with respect to service life when using RA need to be quantified in more detail (e.g. SBS content and fatigue performance, stiffness of asphalt versus low temperature cracking). Issues related to full-scale application (e.g. to avoid degradation of the polymer at high temperatures, mixing time influence on homogeneity and performance of asphalt) also need to be explored. And finally quality assessment and assurance of the RA has to be managed including the quality of the aggregates used. All three challenges can be met through research based on a series of pilot projects. These pilot projects are preferably realized in different European countries to incorporate different recycling techniques and different environmental conditions.

29.9. Acknowledgments

The authors are grateful for the financial support of the trans-national joint research programme "Design – Rapid and Durable Maintenance Methods and Techniques", which is cross-border funded by the Road Authorities of Belgium, Germany, Denmark, Finland, France, Netherlands, Norway, Sweden, Slovenia and the UK. The research programme was initiated by ERA-NET ROAD II.

29.10. Bibliography

[KOM 13] KOMAČKA J., REMIŠOVÁ E., SCHLOSSER F., Asphalt mixtures using reclaimed asphalt containing polymer modified binder, WP4 Report of RECYPMA Project, 2013.

[KOM 14] KOMAČKA J., REMIŠOVÁ E., LIU G. *et al.* "Influence of reclaimed asphalt with polymer modified binder on properties of different asphalts for a wearing course", *3rd International Conference on Transportation Infrastructures*, Pisa, Italy, pp. 179-185, 22-25 April 2014.

[LEE 13] LEEGWATER G., VAN GIJLSWIJK R., Environmental and economic benefits of the use of polymer modified reclaimed asphalt, WP5 Report of RECYPMA Project, 2013.

[LIU 13a] LIU G., VAN DE VEN M., LEEGWATER G., Properties of aged polymer modified binder, WP3 Report of RECYPMA Project, 2013.

[LIU 13b] LIU G. *et al.* "Evaluating the rheological properties of PMB-containing RA binders from surface-layer asphalt mixtures to be recycled", *Construction and Building Materials*, vol. 49, pp. 8–14, 2013.

[NIE 13a] NIELSEN E. State of the art – recycling polymer modified asphalt, WP2 Report of RECYPMA Project, 2013.

[NIE 13b] NIELSEN E., RAABERG J., NEIDEL A., Microscopy analysis of asphalt mixtures using reclaimed asphalt containing polymer modified binder, WP4 Report of RECYPMA Project, 2013.

[VAN 09] VAN DEN BERGH W., VAN DE VEN M., "Hot recycling of asphalt: experiences in Flanders and the Netherlands", *The 2nd International Conference on Environmentally Friendly Roads*, Warsaw, Poland, 2009.

30

Case Study: Increasing the Percentage of Recycled Asphalt

This chapter presents a case study to increase the percentage of reclaimed asphalt (RA) in asphalt mixture by using rejuvenator and to characterize the quality of such a mixture. The effect of rejuvenator was determined on laboratory- and plant-prepared samples. In laboratory, we prepared asphalt mixtures containing different percentages of RA (0, 10, 30 and 50%) and rejuvenator. The mixture without RA is control mixture, made only with virgin raw materials. In asphalt plant, we prepared asphalt mixture with highest content of RA and control mixture. We conducted different tests, e.g. determining softening point, Fraass breaking point, penetration and indirect tensile strength. Asphalt containing RA was successfully produced in ordinary batch asphalt plant with no special hardware, such as parallel drum. Results showed that asphalt mixture containing RA and rejuvenator fulfilled the technical standards, and the quality was in most cases even better than asphalt mixture made from virgin materials.

30.1. Introduction

Asphalt mixtures are commonly used for surface layers on European roads. Asphalt is a mixture of stone aggregate, binder, which is usually bitumen, and air voids. When the construction of new roads or the reconstruction of old roads is in progress, we can choose to use virgin material or we can reuse materials from used asphalt, because asphalt is a material that can be almost entirely recycled. Recycling is the process in which reclaimed asphalt (RA) materials substitute virgin materials. At first, asphalt recycling was considered as an option in the 1970s during the oil crisis, when the prices of bitumen greatly increased. Initially, recycled asphalt materials were primarily used for road maintenance and construction of low-trafficked roads. Today, recycling methods are also considered for heavy trafficked

Chapter written by Marjan TUŠAR and Lidija AVSENIK.

roads, and the recycling ratio may be close to 100% [KAR 06, OLI 13]. We need to take into account that during road service lifetime, the old bitumen has been aged and the aggregate degraded. Also the reclamation process may contribute to the inhomogeneity of the recycled pavement material [KAR 06]. In general, bitumen is subjected to aging in two stages: short-term and long-term aging. Short-term aging occurs during the mixing and construction process of asphalt; long-term aging continues throughout the lifetime of the pavement [LU 01]. Different factors influence the aging process; the most important among them are high temperature and ultraviolet (UV) radiation. The aging process can be divided into two types. The main aging mechanism is an irreversible one, characterized by chemical changes in the binder. These processes are oxidation, loss of volatile components and exudation. The second mechanism is the reversible process called physical hardening [LU 01]. The most important aging process is oxidation of bitumen. The result of oxidation is stiffening or hardening of the bitumen [BRA 93]. It was found that two parallel reactions occurred during the oxidation: oxidation of base asphalts and degradation of polymers [RUA 06]. Because of oxidation, polar functional groups arise, which increase the viscosity by increasing the intermolecular interactions [KAR 06]. As asphalt binder reacts and loses some of its components during the aging process, its rheological behavior will naturally differ from virgin materials [ALQ 07]. Because of aging, bitumen becomes harder, more brittle, its viscosity increases which results in loss of adhesion, ductility and in the end chip loss and cracking of asphalt layers [SAN 05]. An additive called a rejuvenator has to be added to asphalt made from recycled material, so that asphalt can attain the required technical standards. The purpose of rejuvenator is to return aged bitumen its original properties by reconstructing the chemical composition of the aged bitumen [SHE 07]. Rejuvenators are designed to soften the existing bitumen on surface of the RA. They are often included in cold mixture and hot mixture recycling processes. Commonly bitumen becomes less viscous, more ductile and its coating properties are restored [BRO 10]. Due to different step of technological development and public awareness, asphalt recycling rate in Europe vary from 90% to less than 10% [EAP 12]. Asphalt recycling is a demanding task that requires extra knowledge and experience compared to the production of asphalt from virgin materials [KAR 06]. In Slovenia, only some asphalt plants recycle asphalt. The asphalt plants that use RA are not equipped with additional technology such as a parallel drum enabling usage of more than 30% RA in a new asphalt mixture. Our experimental study was oriented to increase RA percentage in new asphalt with minor adaptations of existing technology.

30.2. Materials and test program

The case study consisted of two parts. In the first part, we prepared asphalt samples in the laboratory, and in the second part, asphalt samples were prepared in the asphalt plant. In the laboratory, we prepared and tested samples with different shares of RA (from 0–50%). In the asphalt plant, we prepared only samples with the highest amount of RA (almost 50%) and control mixture without RA.

30.2.1. *Materials*

In the laboratory, we prepared 7 kg of each mixture with 5% of bitumen content (by mass). The RA and rejuvenator were not heated and were added as the last component in laboratory mixer. For preparation of laboratory asphalt samples, we used RA with the properties listed in Table 30.1. The characteristics of RA such as gradation, type and properties of the residual bitumen play a fundamental role in the mix design phase. Other characteristics such as homogeneity, foreign matters and water content significantly influence the quality of the production [GRI 13].

Sieving curve	Unit	
0.063 mm	% by mass	11.0
0.25 mm	% by mass	16.5
0.71 mm	% by mass	25.8
2 mm	% by mass	44.2
4 mm	% by mass	60.8
8 mm	% by mass	79.9
11 mm	% by mass	91.0
16 mm	% by mass	98.8
Content of bitumen	% by mass	3.8
Softening point	°C	78
Penetration	1/10 mm	11.8

Table 30.1. *Properties of RA used in experimental study*

The composition of laboratory-prepared asphalt samples are presented in Table 30.2.

In the asphalt plant, we prepared an asphalt mixture containing 48.5% RA and rejuvenator and control mixture without RA as presented in Table 30.3.

As a binder in the asphalt mixture, we used conventional bitumen with the following properties listed in Table 30.4.

RA (% by mass)	Unit	0	10	30	50	30 without rejuvenator
RA	Kg	0.0	0.701	2.106	3.5	2.106
0.063	% by mass	12.3	12.07	11.05	9.5	11.05
0.25	% by mass	14.11	13.97	13.24	12.18	13.24
0.71 mm	% by mass	20.0	19.45	17.76	14.86	17.76
2 mm	% by mass	34.0	32.97	30.13	24.79	30.13
4 mm	% by mass	50.84	51.19	49.05	44.84	49.05
8 mm	% by mass	69.85	70.30	68.26	64.67	68.25
11 mm	% by mass	81.85	82.49	80.91	78.10	80.90
16 mm	% by mass	98.85	100.0	99.98	100.0	99.98
Rejuvenator	kg	0	0.0103	0.0309	0.0515	0.0
Fresh bitumen	kg	0.35	0.3131	0.2393	0.1655	0.27
Bitumen content	% by mass	5.0	5.0	5.0	5.0	5.1

Table 30.2. *Properties of laboratory-prepared asphalt samples*

Sieving curve	Unit	Plant mixture without RA	Plant mixture with RA	Laboratory mixture with RA
0.063 mm	% by mass	8.6	9.5	8.8
0.09 mm	% by mass	9.0	10.0	10.0
0.25 mm	% by mass	13.0	14.0	14.0
0.71 mm	% by mass	21.0	22.0	21.0
2 mm	% by mass	39.0	40.0	35.0
4 mm	% by mass	54.0	56.0	49.0
8 mm	% by mass	76.0	78.0	71.0
11.2 mm	% by mass	92.0	90.0	87.0
16 mm	% by mass	100.0	96.0	100.0
22.4 mm	% by mass	100.0	97.0	100.0

Table 30.3. *Sieving curve of asphalt samples prepared in asphalt plant*

Property	Unit	Value
Penetration	1/10 mm	94
Softening point	°C	46.1
Fraass breaking point	°C	–19.0

Table 30.4. *Properties of used fresh bitumen*

30.2.2. *Test program*

All asphalt samples were prepared in accordance with SIST EN 12697-30:2004+A1:2007. Laboratory asphalt samples contain different percentages of RA (from 0 to 50%) and rejuvenator. All properties of samples were compared to properties of the control mixture, which was prepared only with virgin materials and did not contain any RA. On prepared samples, we performed different tests according to European Standards. Tests (Table 30.5) showed properties of extracted bitumen and properties of asphalt mixtures.

Samples	Test	European Standard
Extracted bitumen	Softening point	SIST EN 1427:2007
	Penetration test	SIST EN 1426:2007
	Fraass breaking point	SIST EN 12593:2007
All samples	Determining void content	SIST EN12697-8:2004
	Bulk density	SIST EN 12697-6:204+A1:2007
	Maximal density	SIST EN 12697-5:2010
	Indirect tensile test	SIST EN 12697-23:2004
	Sample preparation	SIST EN 12697-30:2004+A1:2007
In plant-prepared samples	Marshall test (stability and flow)	SIST EN 12697-34:2004+A1:2007
	Water sensitivity tests	SIST EN 12697-12:2009
	Slab compaction	SIST EN 12697-33:2004+A1:2007
	Uniaxal tension test	prEN 12697-46:2010
	Wheel tracking test	SIST EN 12697-22:2004+A1:2007

Table 30.5. *Tests carried out in research*

On extracted bitumen, we determined the softening point using a ring and ball method (SIST EN 1427:2007). We checked the consistency of bitumen with the standard needle penetration test (SIST EN 1426:2007). The Fraass breaking point was determined in accordance with SIST EN 12593:2007. The Fraass breaking point is criterion for brittleness of bitumen at low temperatures.

We gained basic information about asphalt samples by determining void content (SIST EN 12697-8:2004), bulk density (SIST EN 12697-6:204+A1:2007), maximal density (SIST EN 12697-5:2010) and indirect tensile test (SIST EN 12697-23:2004). We conducted more tests on asphalt samples prepared in the asphalt plant. We used

Marshall test to determine stability and flow (SIST EN 12697-34:2004+A1:2007). Water sensitivity tests were made according to SIST EN 12697-12:2009 to evaluate the effect of moister. Asphalt mixture produced in the asphalt plant was compacted in the form of slab (SIST EN 12697-33:2004+A1:2007) from which specimens were cut out. On these specimens, we carried out a uniaxial tension test (prEN 12697-46:2010) for characterizing the resistance against low temperature cracking.

An asphalt mixture prepared in the asphalt plant was also laid on the test field. A week after paving, we took samples and determined bulk density, void content and preformed the wheel tracking test (SIST EN 12697-22:2004+A1:2007). After 1 year, we took other samples from the test field and repeated the tests.

30.3. Results and analysis

Results for asphalt mixtures prepared in the laboratory are shown in Table 30.6, and results for asphalt mixtures prepared in the asphalt plant are shown in Tables 30.7 and 30.8.

During the production of asphalt mixture in the laboratory, we had to prolong the mixing time to get suitable homogeneous mixtures.

RA (% by mass)	Unit	0	10	30	50	30 without rejuvenator
Softening point	°C	48.4	50.6	63.2	75.4	53.4
Penetration	1/10 mm	65.0	63.0	63.0	54.0	44.0
Fraass breaking point	°C	–11.0	–12.0	–15.0	–18.0	–11.0
ITS (2 ×50 blows) at 25°C	MPa	0.93	0.97	0.91	0.91	1.23
Temperature of compaction	°C	150.0	155.0	160.0	165.0	160.0

Table 30.6. *Characteristics of extracted bitumen from laboratory-prepared asphalt samples*

From the producers' declarations, we found that rejuvenators consisted of soft bitumen, special oil and paraffin wax. It was expected that oil and soft bitumen would lower viscosity of fresh bitumen and melt some aged bitumen in RA. The function of paraffin wax was to increase the viscosity of the bitumen at lower temperatures, thus increasing the resistance to permanent deformation immediately after the laydown of asphalt pavement. Results showed that the softening point increased with increasing percentage of RA and rejuvenator. The softening point of bitumen extracted from the control mixture was 48.4°C and of bitumen extracted from the asphalt mixture containing 50% RA was 75.4°C. By observing the Fraass breaking point, we found that its value decreased with the addition of RA and rejuvenator. At first (for bitumen from control mixture), it was −11°C and for bitumen from asphalt mixture with highest amount of RA, it was −18°C. Increasing the softening point and at the same time decreasing the Fraass breaking point means that the service temperature of bitumen increased. When only RA was added and no rejuvenator, the softening point increased compared to control mixture, but when we compare it to mixture containing the same amount of RA (30%) and a rejuvenator, we found a decrease in the value of softening point. In the last sample, 30% of RA and no rejuvenator, the Fraass breaking point has the same value as measured for bitumen from the control mixture. The penetration value decreased, when we added RA, from 6.5 mm for control mixture to 5.4 mm for the sample with 50% RA. The results show a smaller reduction of penetration when rejuvenator was added. Mechanical properties of asphalt mixtures tested according to EN 12697-23 (indirect tensile test) were almost uniform for all mixtures even when rejuvenator was added. Other studies report contradictory results about ITS [SHU 07, SU 09, VAL 10]. The compaction temperature increased when we added RA. Comparing samples with RA and rejuvenator to the samples with RA and no rejuvenator, it has been found that the compaction temperature was the same.

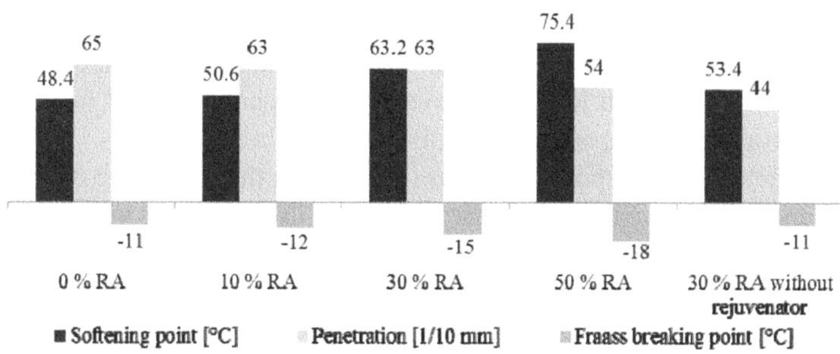

Figure 30.1. *Characteristics of extracted bitumen from laboratory-prepared asphalt mixtures with different percentages of RA and with or without rejuvenator*

In the process of preparation of asphalt mixtures in the asphalt plant, RA was continuously added at the end of the dryer drum, with dosing conducted visually by setting a conveyer belt speed. At such production, RA was heated indirectly with a mineral mixture in hot elevator. In order to ensure a proper mixing temperature, the mineral mixture had to be heated to higher temperatures. For such a high proportion of RA (50%), the temperature of stone aggregate mixture should be increased for at least 100°C (to at least 250°C), but our technology did not allow us to carry out the process at this temperature. With the additive used, production temperature can be lower. By measuring the mass of RA entering the production system, we provided

50% of RA at the first phase dosing. The mineral mixture in the drum dryer was heated to 220–230°C and the production temperature of the asphalt mixture in a mixing drum was 110–150°C. Due to the small stone aggregate mass flow through the dryer drum in these conditions of production, fume temperatures were too high. To ensure proper operation of the asphalt plant, it was necessary to slightly decrease RA content. Estimated share of RA in new asphalt mixtures in steady-state conditions was 40–45%. In addition to production temperature, correction of the mixing time was required to provide an even distribution of additive in asphalt mixtures. As a rule of thumb, the higher the RA temperature before mixing, the higher the amount of RA to be recycled. If cold and wet RA is added directly in the mixing drum, the virgin aggregates have to be superheated in order to ensure an appropriate mixing temperature. The excessive temperature of aggregates can cause oxidation, or even burning, of the bitumen that can compromise the durability of the final asphalt mixture [GRI 13]. Mixing time was prolonged from 40 to 48 sec. We were not able to prolong the mixing time for more than 8 sec due to software limitations. In the produced asphalt mixture, we still observed some unmixed granules of RA.

Results on asphalt mixtures prepared in the asphalt plant (Table 30.7) confirmed the results from the laboratory. The softening point of extracted bitumen increased when we added rejuvenator. After comparing softening point of extracted bitumen from asphalt mixture with RA and rejuvenator produced in the laboratory and in plant, we can see they are similar, but the penetration values are more similar between plant mixture with RA and rejuvenator compared to plant mixture without RA. Volumetric properties are comparable for all three mixtures. We conducted mechanical tests only on asphalt mixtures produced in the plant. The conclusions are that mixtures with RA and rejuvenator had better low temperature properties. This was confirmed with the uniaxal test, i.e. thermal stress restrained specimen test (TSRST) method, which showed a lower temperature before the failure of the specimen with RA and rejuvenator comparing to control specimen. Mixtures with RA and rejuvenator were less resistant to compaction, nonetheless they were less sensitive to water than the control mixture.

At the test field, we laid asphalt mixtures with RA and rejuvenator and control mixture without RA. An asphalt mixture without RA was built in at temperature around 170°C, and the mixture prepared with RA and additive was built in at temperature around 100°C, but was also built manually at 70°C. Compaction of both asphalt mixtures on the test field took place without problems. The asphalt mixture containing RA was compacted even a bit easier. The realization of joints and manual work was also better and easier. For determining long-term behavior, we performed the wheel tracking test on mixtures built-in test field. One week after production, the wheel tracking test showed better properties of control mixture, but a year after, the results were opposite, asphalt containing RA and rejuvenator was more resistant to rutting (Table 30.8 and Figure 30.2). All asphalt properties are within acceptable limits. From Table 30.7, it can be also seen that void content on the test field is higher than in the laboratory.

Property	Unit	Plant mixture without RA	Plant mixture with RA	Laboratory mixture with RA
Bitumen content	% by mass	5.0	4.9	5.1
Softening point	°C	52.5	77.6	75.4
Penetration	1/10 mm	42	42	54
Void content	% by volume	2.2	1.8	1.9
Voids filled with bitumen	% by volume	84.6	86.8	86.9
Bulk density	kg/m^3	2,462.6	2,476.3	2,485.0
Maximal density	kg/m^3	2,517.5	2,521.4	2,532.0
Marshal stability at 60°C	kN	10.4	11.0	–
Marshal flow at 60°C	mm	2.6	3.3	–
Marshal stiffness at 60°C	kN/mm	3.9	3.4	–
ITS_d (at 25°C)	kPa	929.5	769.1	910 (2 ×50 blows)
ITS_w (at 25°C)	kPa	888.3	773.6	–
Water sensitivity ITSR	%	95.6	100.6	–
Compactibility C/0.0202 (at 148°C)	21 Nm	30.7	24.5	–
Failure stress $\sigma_{cry.\ failure}$	MPa	3.9	5.2	–
Failure temperature $T_{failure}$	°C	–27.5	–32.0	–

Table 30.7. *Properties of asphalt samples without RA and asphalt samples from the plant or laboratory containing 50% RA and rejuvenator*

Time after paving/sample/ temperature	Bulk density	Thickness	Void content	Compaction degree	Average rut depth	Proportional rut depth PRD_{AIR}	Rate of rut propagation WTA_{AIR}
	(kg/m^3)	(mm)	(%)	(%)	(mm)	(%)	(mm/1,000)
1 Week/with RA and rejuvenator/55°C	2,388	62	5.1	97.0	6.22	10.0	0.28
1 Week/without RA and rejuvenator/55°C	2,407	60	4.5	97.2	4.17	7.0	0.13
1 Week/with RA and rejuvenator/50°C	2,390	67	5.1	97.1	5.9	8.8	0.21
1 Week/without RA and rejuvenator/50°C	2,418	68	4.1	97.6	5.05	7.4	0.1
1 Year/with RA and rejuvenator/55°C	2,394	76	4.9	97.2	1.84	2.4	0.08
1 Year/without RA and rejuvenator/55°C	2,418	63	4.1	97.6	3.98	6.3	0.2
1 Year/with RA and rejuvenator/50°C	2,381	74	5.4	96.7	1.02	1.4	0.04
1 Year/without RA and rejuvenator/50°C	2,417	63	4.1	97.6	2.96	4.7	0.1

Table 30.8. *Properties of built-in asphalt course with or without RA*

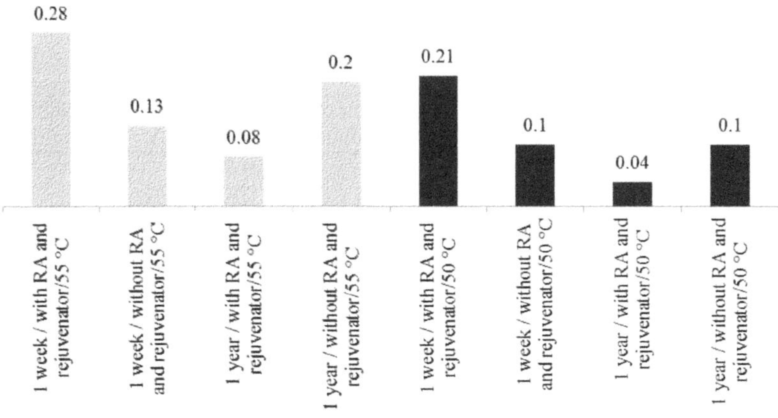

Figure 30.2. *Average rut propagation (mm/1,000) of asphalt taken from the test field*

30.4. Conclusions

From the results of this experimental study, we can conclude the following: the amount of RA can be increased by using rejuvenator and the quality of such asphalt mixture is in most cases equal to or even better than asphalt mixture made up of virgin materials. The use of rejuvenator is reasonable even for low RA content because rejuvenator improves the properties of asphalt mixture. The use of RA and rejuvenators is cost effective, but it must be a daily practice, because if we use RA only occasionally, then such productions represent a disturbance in regular production. The most expensive component of asphalt mixture is bitumen and by using RA and rejuvenator we can lower the content of fresh bitumen (Table 30.2). Rejuvenator enables reuse of RA and at the same time improves the properties of asphalt mixtures containing RA. By using rejuvenator, we can lower the compaction temperature, which is cost efficient, saves energy and is more ecological. Lower compaction temperatures means longer transportation distance and time [MER 11]. Rejuvenator prolongs the life time of the road surface [KAR 06]. Much lower temperatures also mean much lower energy consumption in the production of asphalt mixtures, and thus reduced greenhouse gas emissions. When installing asphalt, it means less emission of bitumen fumes and vapors, and is thus less harmful to the health of workers. Equally important is that by using RA, we preserve nature, reduce usage of virgin raw materials and reduce waste.

In our study, we showed that an asphalt mixture prepared in the asphalt plant containing about 50% RA was successfully produced in an ordinary batch asphalt plant with no special hardware for RA addition such as parallel drum. In spite of low temperatures after production (about 100°C), the asphalt mixture was also successfully laid and compacted as a wearing course in a test section. Asphalt mixtures with RA and rejuvenator showed better test results.

30.5. Bibliography

[ALQ 07] AL-QADI I.L., ELSEIFI M., CARPENTER S.H., "Reclaimed Asphalt pavement – a literature review", *Research Report FHWA-ICT-07-001*, Illinois Center for Transportation, 2007.

[BRA 93] BRANTHAVER J.F., PETERSEN J.C., ROBERTSON R.E., *et al.*, "Binder Characterization and evaluation, Volume 2: Chemistry", *Strategic Highway Research Program*, National Research Council, Washington, DC, 1993.

[BRO 10] BROWNRIDGE J., "The role of an asphalt rejuvenator in pavement preservation: use and need for asphalt rejuvenation", *Compendium of Papers from the First International Conference on Pavement Preservation*, University of California, Berkeley, CA, pp. 351–364, 13–15 April 2010.

[EAP 12] EAPA, Asphalt in figures, 2012, available at http://www.eapa.org/asphalt.php, 2012.

[GRI 13] GRILLI A., BOCCI M., CARDONE F. *et al.*, "Laboratory and in-plant validation of hot mix recycling using a rejuvenator", *International Journal of Pavement Research and Technology*, vol. 6, no. 4, pp. 364–371, 2013.

[KAR 06] KARLSSON R., ISACSSON U., "Material-related aspects of asphalt recycling – state-of-the-art", *Journal of Materials in Civil Engineering*, vol. 18, no. 1, pp. 81–92, 2006.

[LU 01] LU X., ISACSSON U., "Effect of aging on bitumen chemistry and rheology", *Construction and Building Materials*, vol. 16, no. 1, pp. 15–22, 2001.

[MER 11] MERUSI F., GIULIANI F., "Rheological characterization of wax-modified asphalt binders at high service temperatures", *Materials and Structures*, vol. 44, no. 10, pp. 1809–1820, 2011.

[OLI 13] OLIVEIRA J.R.M., SILVA H.M.R.D., JESUS C.M.G. *et al.*, "Pushing the asphalt recycling technology to the limit", *International Journal of Pavement Research and Technology*, vol. 6, no. 2, pp. 109–116, 2013.

[RUA 03] RUAN Y., DAVISON R.R., GLOVER C.J., "Oxidation and viscosity hardening of polymer-modified asphalts", *Energy & Fuels*, vol. 17, pp. 991–998, 2003.

[SAN 05] SANDERS P., Silvia project report. Review of recycling and rejuvenation procedures. Sustainable road surfaces for traffic noise control, European Commission DG Tren – Growth, contract GRD2-2000-31801-SI2.335701, p. 34, 2005.

[SHE 07] SHEN J., AMIRKHANIAN S., MILLER J.A., "Effects of rejuvenating agents on superpave mixtures containing reclaimed asphalt pavement", *Journal of Materials in Civil Engineering*, vol. 19, no. 5, pp. 376–384, 2007.

[SHU 07] SHU X., HUANG B., VUKOSAVLJEVIC D., "Laboratory evaluation of fatigue characteristics of recycled asphalt mixture", *Construction and Building Materials*, vol. 22, pp. 1323–1330, 2007.

[SU 09] SU K., HACHIYA Y., MAEKAWA R., "Effect of recycling rate and asphalt binder on the performance of recycled asphalt concrete for airport pavement rehabilitation", *Road Materials and Pavement Design*, vol. 10, no. 2, pp. 361–371, 2009.

[VAL 10] VALDES G., PEREZ-JIMENEZ F., MIRO R. *et al.*, "Experimental study of recycled asphalt mixtures with high percentages of reclaimed asphalt pavement", *Construction and Building Materials*, vol. 25, pp. 1289–1297, 2010.

31

Evaluation of Long-term Glass-grid Test Section using a Unique Method

The objective of this study was to evaluate a glass-grid-reinforced test section after 12 years of service. A test section was established in 2001 on the street which was subjected to the deep reconstruction. The total length of the test section was approximately 700 m and was divided into two subsections, with and without the glass-grid installed between asphalt sub-base and asphalt binder course. Effectiveness of the grid reinforcement was evaluated in 2004, 2008 and 2012. The data evaluation included a maximum deflection comparison as well as a deflection bowl geometry analysis. The geometry analysis was based on the hypothetical shape of the deflection bowl, that was virtually divided into the arch segments between three consecutive points. Reinforced sections were characterized by greater arch radiuses than on the control section. It was recognized as the measurable reinforcing effect of the glass-grid on the pavement fatigue life. Maximum deflection growth was higher on the unreinforced section and more deterioration was observed on the unreinforced section.

Nowadays, a falling weight deflectometer (FWD) is the most common non-destructive device for bearing capacity assessment. Different pavement types are evaluated during and after the construction, such as typical roads and city streets, airport pavements, and large industrial surfaces. The analysis of load capacity is made on the basis of deflection bowl measurements. The pavement deflection caused by the load is measured with the use of sensors (geophones), which are situated in the pressure center (plate bearing device) and at certain distances from the central point (e.g. 300, 600, 900, 1,200, 1,500 and 1,800 mm). Knowing the

---

Chapter written by Maciej MALISZEWSKI, Przemysław HARASIM, Dominika MALISZEWSKA and Adam ZOFKA.

pavement construction and the depth and shape of the deflection bowl, the pavement condition can be assessed. During routine measurement, the FWD device simulates the pressure generated by a truck moving at a speed of 35–40 km/h, but it has the ability to apply a maximum force of 120 kN.

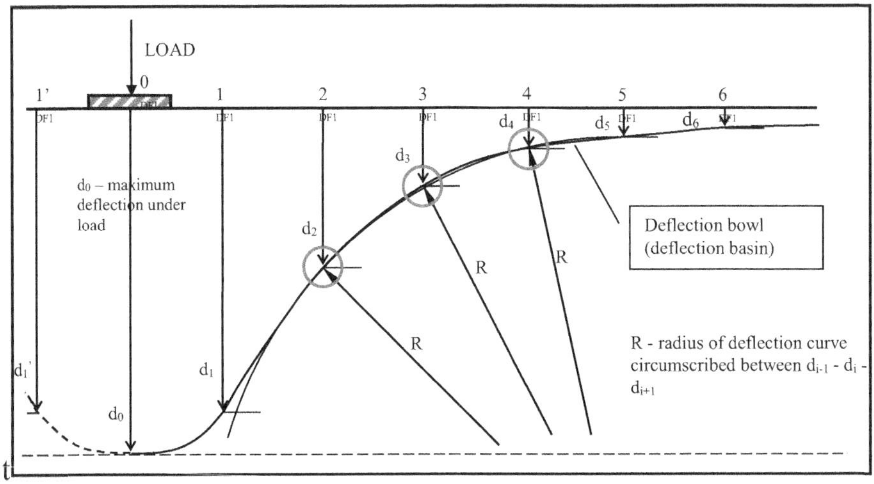

Figure 31.1. *Definition of terms*

31.1. Introduction

31.1.1. *General*

The variety of geosynthetics for the reinforcement of asphalt layers makes it difficult to choose an optimal method for the assessment of their effectiveness. The reinforcing capabilities are affected by product type (grid, composite, nonwoven), raw materials (mineral fibers, polymers, steel rods, etc.) as well as by the surface coating. Perhaps the most important is the quality of application process and product location in a pavement construction. The afore-mentioned characteristics contrast the development of the universal design method of asphalt layers reinforced using geosynthetics. The only available method of assessment is the indirect, instrumented evaluation of such constructions [BAŃ 12]. This chapter describes a specific assessment method of the long-term influence of the glass geogrid on the asphalt pavement. This study was partly funded within the scope of the SPENS Project (6th Framework Program) [MAL 09] and the Road and Bridge Research Institute (IBDiM) research budget.

31.1.2. *Geosynthetic*

In Polish road construction practice [ZAW 04, SYB 11], the main goal of geosynthetic application between asphalt layers is to mitigate the formation of fatigue cracking. This treatment is mainly applied in pavements consisting of flexible sub-base and/or weak subgrade when there is a high risk of deflections exceeding the threshold levels. If the rigid sub-base is present, geosynthetic application is intended to reduce the propagation of reflective cracking. In both cases, geosynthetics play their role only if they are placed in tension stress areas while the pavement is loaded by wheel or sub-base shrinkage occurs. In order to fulfill these objectives, the following conditions should be met:

– location of geosynthetics as close to the base of the asphalt layers as possible (not always possible in case of overlaying old pavements);

– the use of geosynthetics characterized by a stiffness modulus higher than the surrounding asphalt layers throughout the temperature range;

– good geosynthetic bonding with surrounding asphalt layers;

– pre-tensioned placing of geosynthetics.

At the design stage, as well as during the construction process, these requirements are generally taken into account. However, the real results of the aforementioned assumptions can be only proven on the basis of long-term observations or "in situ" test results, which is the main focus of this chapter.

The glass grid is one of many types of reinforcement used in road construction. Its functions are mainly to improve pavement fatigue life, rutting resistance and reflective cracking propagation. In this project, the pavement was reinforced with a grid made of glass fibers (a so called glassgrid). The grid was characterized by 35 kN/m ultimate strength and less than 3% ultimate strain. Nowadays, glassgrids are produced to reach higher strength levels, but in 2001 it was a standard type.

31.1.3. *Test section*

The test section was established on eastern carriageway of Marszałkowska Street in Warsaw. Marszałkowska Street was deep reconstructed in 2001, between crossings with Piękna Street and Nowogrodzka Street. Old pavement was overhauled and replaced with the flexible pavement construction, which is presented in Table 31.1.

Layer/course	Material	Thickness (cm)
Wearing course	AC 16 PmB	5
Binder course	AC 20 PmB	5
Reinforcing grid*	Glass fiber	ca. 0.5
Sub-base	AC 32 35/50	9
Sub-base	Mechanical stabilization	25
Anti-frost layer	Allin aggregate	20
Non-woven*	Polypropylene	ca. 0.5
Subgrade	Aggregate	10 cm

*Present only on half of test section – *the reinforced section.*

Table 31.1. *Typical pavement section on Marszałkowska Street*

Every layer above the granular sub-base was tack coated using asphalt cationic emulsion. The total length of the reconstruction was approximately 700 m. It was divided into two subsections, each 350 m long. The primary subsection between Piękna Street and Hoża Street contained the installation of a glass-grid between asphalt sub-base and asphalt binder course and non-woven on top of subgrade. Accordingly, that subsection is called the *reinforced section* in this chapter, whereas the second section, between Hoża Street and Nowogrodzka Street, will be named the *unreinforced section*.

31.2. Methodology

31.2.1. *Theory and assumptions*

The main problem of the evaluation of asphalt layer reinforcement effectiveness in the first years after construction is a lack of visual performance indicators. More advanced methods should be used. Theoretical models predict the behavior of the pavement, but must be calibrated [JEM 08]. Direct type measurements using stress and strain gauges are more accurate to evaluate pavement condition, but are more time- and cost-consuming compared to indirect pavement condition evaluation methods.

In this research project, the reinforcing effect of glass-grid application was evaluated using the FWD. The FWD produces test results comprising applied stress conditions as well as pavement response in terms of the shape of the deflection bowl. Two types of data were evaluated – maximum deflections and deflection bowl geometry together with corresponding indices. Maximum deflections under the load plate indicate the general pavement condition and remaining fatigue life, expressed in equivalent axle loadings. The higher the recorded maximum pavement deflections, the less load axles this pavement can bear.

Initial analysis showed nearly no influence of reinforcement on the maximum deflections of the pavement, indicating statistically equal mean deflections under the load plate on both reinforced and unreinforced sections. While positive reinforcement effects are recognized and proved by observations, there is almost no evidence of their positive influence confirmed by direct tests performed on real reinforced pavement. The misleading test results were usually explained as being caused by inhomogeneous subgrade or construction, improper installation of reinforcement or errors during measurement. This was the motivation for applying a more advanced approach in order to evaluate the pavement reinforcement effect.

The research was intended to verify if the application of the grid on the reinforced section resulted in a change in maximum deflections under load plate, deflection curve shape and how these parameters were changing in time. The assumption was that when comparing a pair of single FWD measurements indicating the same maximum deflections, the reinforced pavement deflection curvature should have greater radius. It applies to both the horizontal plane or vertical section (Figure 31.2). In order to simplify the comparison, the cross-section of the deflection bowl was divided into the segments comprising three consecutive points (measurements). The coordinates (span and deflection value) of these measurement points were known, so the arch of known radius could be circumscribed between them. In further considerations if the arch was named as d_1–d_3, it means that it was circumscribed between measurements taken from geophone 1–3. The only exception was made in case of arch d_2–$d_{2'}$, where the arch is circumscribed between geophone 2, 1 and 2 mirrored by 1 (that is why it was called 2′). This notation helped to compare both reinforced and unreinforced sections by radius values.

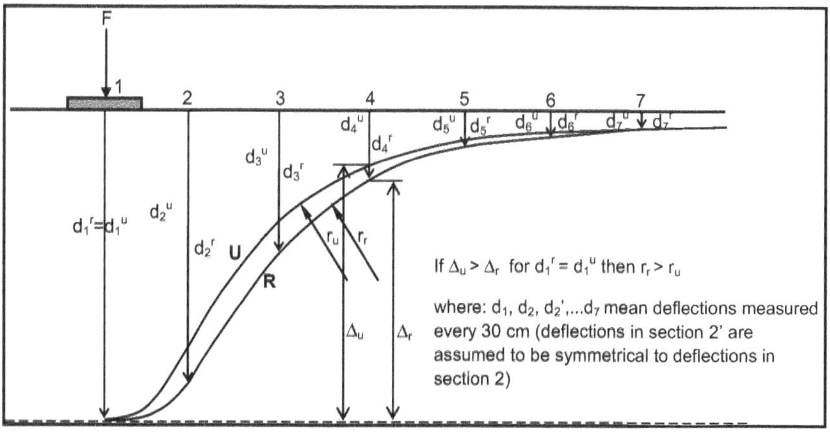

Figure 31.2. *Reinforced/unreinforced deflection curve (illustration of hypothesis)*

31.2.2. *FWD measurement campaign*

The first FWD measurements took place in autumn 2004 and were repeated in [ZAW 07], as well as in 2012 during the same season before seasonal precipitation. Deflection measurements were performed on both reinforced and unreinforced subsections and on one every out of three lanes with a spacing of 25 m. There were two test loads of 50 and 80 kN applied to obtain a wider deflection spectrum. All measurements were performed during one evening. In total, 168 measurements were achieved in each campaign (3 lanes, 2 loads, 28 stations). Single measurement data comprised actual contact pressure, load value and deflection values under every one of seven geophones (2004) or nine geophones (2008 and 2012).

31.2.3. *Visual pavement description*

Pavement condition was assessed in 2008 using the standard visual method according to [SOS 02]. Pavement condition was inspected visually and all deteriorations were recorded. In 2012, the automated pavement distress collection technique was utilized. A system comprising cameras and optical detectors recorded pavement surface, and then deteriorations were automatically detected and categorized by the appropriate software.

31.3. Analysis

31.3.1. *Presentation of deflection basin*

Characteristics of deflection basins on test sections are presented in Figure 31.3. This figure shows the difference between 2004 and 2012 measurements in terms of deflection distribution along section chainage and in offset from load plate. The predicted difference in deflection evolution was confirmed by these experimental data. The first measurements in 2004 were characterized by lower values, particularly on the unreinforced subsection. The last measurements in 2012 showed a higher increase in deflections on unreinforced subsection (indicated by the red areas in Figure 31.3).

31.3.2. *Comparison of maximum deflections*

Geosynthetics built-in between the asphalt layers will reduce fatigue cracking of the pavement, if it reduces maximum deflections under wheel load and/or enlarges the curvature radius when compared with the unreinforced pavement. The 2004 deflection results showed some differences between sections but they were not statistically significant. The difference in maximum deflections between reinforced and unreinforced pavement became significant in the 2008 and 2012 measurement campaigns. Additional observations showed that deflection increase was higher on lane 1 (right-hand lane), which is typically subjected to heavier traffic loads. The comparison charts are shown in Figure 31.4.

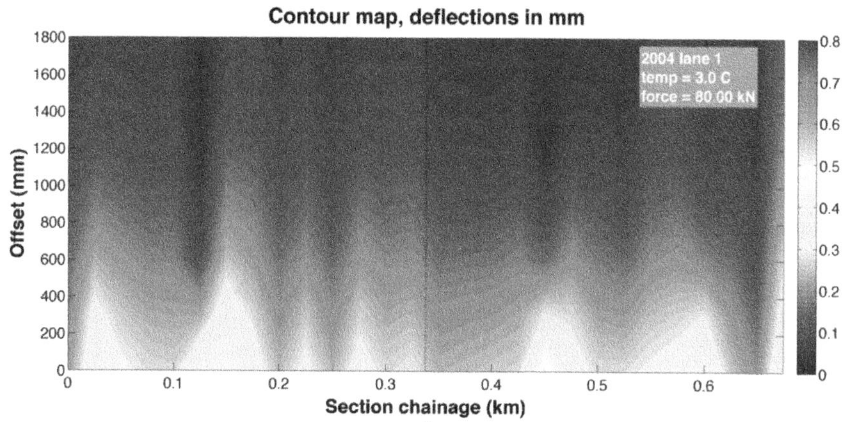

a)

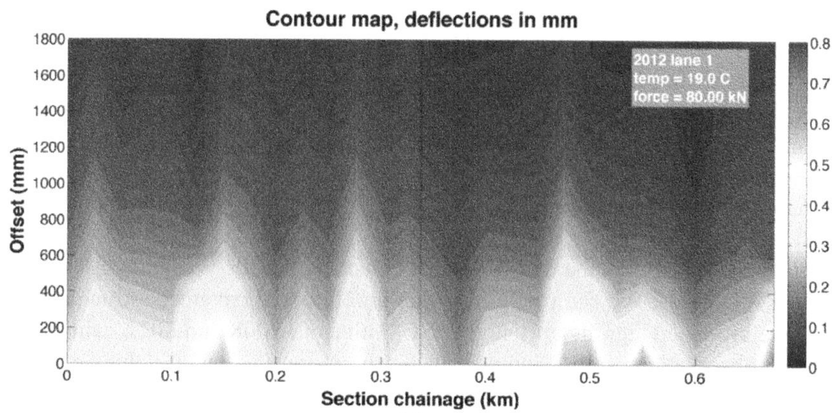

b)

Figure 31.3. *Experimental data from pavement deflection basin: a) 2004 and b) 2012 (reinforced section is between 0 and 0.350 km). For a color version of this figure, see www.iste.co.uk/jacob/materials.zip*

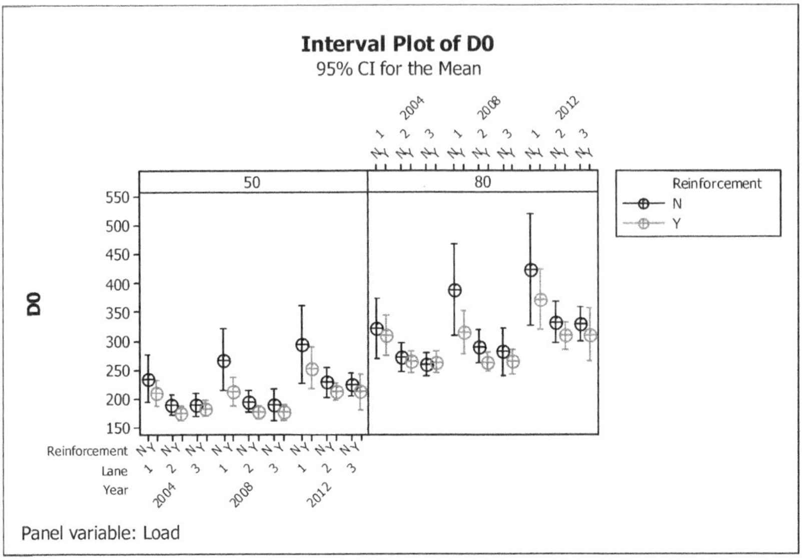

Figure 31.4. *Comparison of maximum deflections in time and location*

31.3.3. *Deflection basin indices*

In addition to the raw deflection values, collected FWD data were also analyzed in terms of the deflection basin indices: surface curvature index (SCI) and well-established AREA parameter. The SCI parameter characterizes the condition of the upper (bound) layers, and is calculated as the difference between deflections at 0 and 300 mm. On the other hand, the AREA parameter uses deflections at 0, 300, 600, and 900 mm in order to calculate the approximate basin area that is further normalized by d_0.

Figure 31.5 presents the SCI trends for considered sections separately for 50 and 80 kN FWD loads. We can easily note that unreinforced pavement structures in the first lane produced significantly higher SCI values in 2008 and 2012, which agrees with the observations from Figure 31.4. Furthermore, the actual values of the SCI from 2012 (lane 1) could suggest that the unreinforced pavement structure requires rehabilitation treatment in order to increase its bearing capacity. Similar trends can be observed in the AREA values, as shown in Figure 31.6. While the AREA parameters were approximately the same in 2004, they differ significantly in 2012, clearly indicating the influence of the reinforcement geogrid.

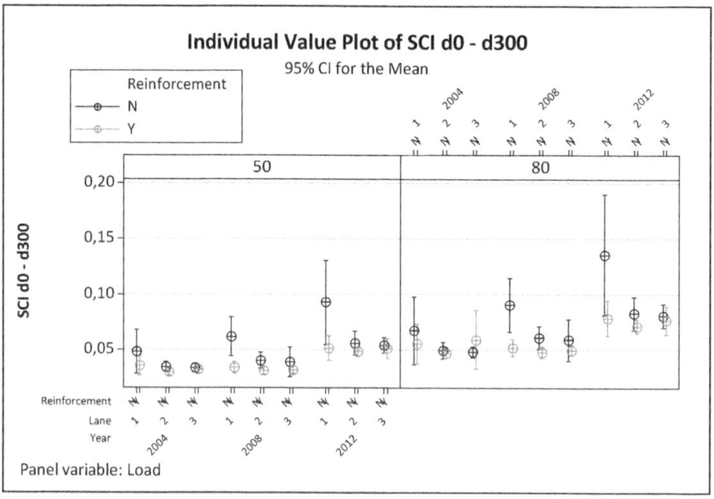

Figure 31.5. *Surface curvature index parameter*

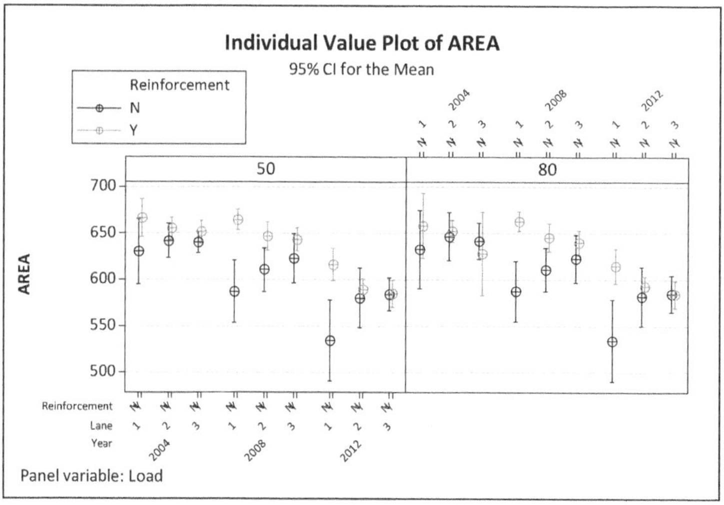

Figure 31.6. *AREA parameter*

31.3.4. *Radius comparison*

The deflection basin was split into eight sections comprising three neighboring points. Results derived for different campaigns and separately on reinforced and unreinforced sections are shown in Figure 31.7. It can be easily concluded that an increase in radius values can be noted in almost every section of the deflection basin. The lowest radius values were noticed in the deflection bowl section located directly under load plate (R-101). This first section had the lowest radius decrement in 8 years of pavement service. Higher average radiuses were located on the reinforced section. The second deflection bowl section (R012) was evaluated only in 2008 and 2012 and radiuses in this section were over two times higher comparing to the first section. Additionally, the second section of the deflection basin encountered a higher decrement of radius in time. The difference between reinforced and unreinforced section was the opposite; higher radiuses were observed on unreinforced section. The third and fourth deflection bowl sections (R123 and R234 accordingly) were similar in evolution. The initial 2004 and 2008 reinforced section radiuses were among highest values of the whole deflection basin. Also the highest radius decrements (with highest confidence level) were observed on the unreinforced section compared to the glass-grid section. In 2012, these two sections of deflection basin showed the highest radius decrement (from over 20,000 m to less than 3,000 m) on both reinforced and unreinforced sections. These sections of deflection curve presented pavement bearing capacity decrease over time. Also fifth (R345) and sixth (R456) deflection basin sections presented decrement of radius over time. In this case, the first 7 years of pavement service were assisted by glass-grid reinforcement, which limited the radius value increment. In 2012, radiuses of reinforced and unreinforced sections equalized in this part of the deflection basin. The seventh (R567) deflection basin section presented initial growth of radius in 2008 and after that dropdown in 2012. Similarly, the reinforced section was characterized by the higher values of radiuses. The last section of deflection curve (R678) showed different behavior; neither radius value increment in time nor reinforcement influence on radius was significant.

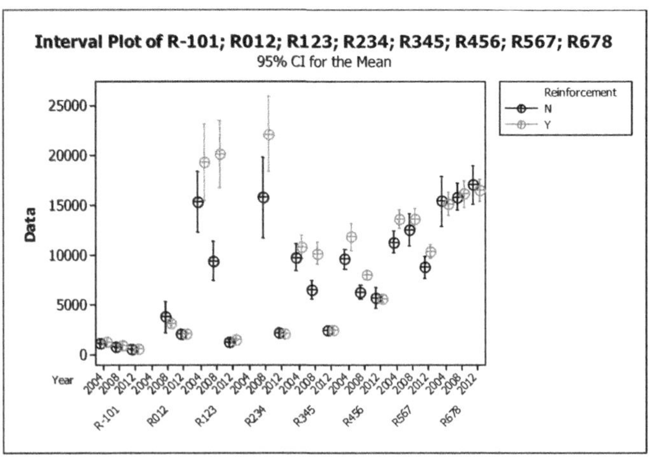

Figure 31.7. *Evolution of deflection basin radiuses over time*

31.3.5. *Visual pavement description*

Pavement inspection results from 2008 and 2012 are described in Table 31.2. In 2008, minor deteriorations were localized. Inspection indicated that reinforced pavement showed less distress than unreinforced, but differences were not significant. In 2012, more deterioration occurred on both sections. In general, deteriorations were not severe. In terms of the bearing capacity, alligator cracks were significant. The majority – two-thirds – of them were located on unreinforced section. It should be pointed out that the reinforcing grid did not protect the pavement against longitudinal and transversal cracks. It must be emphasized that the pavement was already 11 years old during the second visual inspection.

Visual inspection campaign	Deteriorations on reinforced section	Deteriorations on unreinforced section
2008	1 longitudinal crack (1 m)	no longitudinal cracks
	2 alligator cracks (0.5 m^2 total)	3 alligator cracks (1.5 m^2 total)
	2 transversal cracks (1 m total)	4 transversal cracks (11.5 m total)
	1 patched location (4 m^2)	3 patched locations (12 m^2 total)
2012	19 longitudinal cracks (59 m)	7 longitudinal cracks (18 m total)
	14 alligator cracks (14 m^2 total)	21 alligator cracks (21 m^2 total)
	26 transversal cracks (46 m total)	16 transversal cracks (31 m total)
	5 patched location (40 m^2)	12 patched locations (70 m^2 total)

Table 31.2. *Visual inspection results*

Figure 31.8 shows an attempt to examine the relationships between observed distresses and considered FWD parameters (SCI, AREA and R234). The distresses were expressed in terms of the fatigue/block cracking area that was normalized by the highest value recorded among the 25 m segments. Although all trends show weak or a lack of correlation, this analysis should be repeated during the following inspection in order to investigate distress evolution and verify the critical values of the considered FWD parameters.

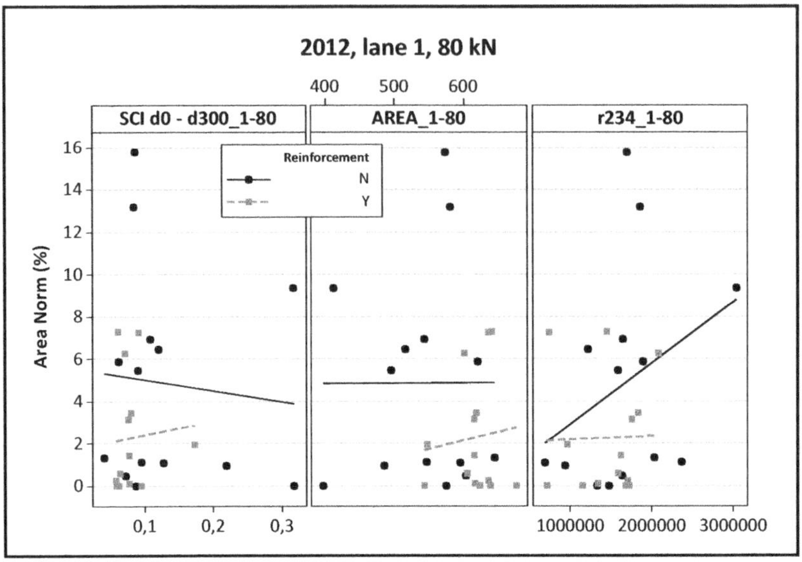

Figure 31.8. *Observed distress (fatigue/block cracking area) versus considered FWD parameters in 2012*

31.4. Conclusions and recommendations

The application of a glass-grid between the asphalt sub-base and binder course in the flexible structure only insignificantly decreased the maximum FWD deflections 4 years after reconstruction. After the following four years (in 2008), the difference between maximum FWD deflections became significant. Also the influence of reinforcement onto the deflection basin radiuses was the most significant. This observation applied mostly to lanes subjected to heavier traffic load. Comparing measurements made on reinforced section in 2004 and 2008, an insignificant change in maximum deflections has occurred. The same comparison made on unreinforced section showed a significant increase in deflections in the case of lane 1, which was subjected to bus traffic, qualified as heavy. It can be pointed out that the influence of reinforcement on deflection value became visible after 8 years under traffic. Reinforcement activates as soon as deterioration factors are present (in this case heavy bus traffic). The application of a glass-grid between the asphalt sub-base and binder course in flexible construction resulted in a beneficial increase indeflection curve arc radiuses on the reinforced section while comparing to the unreinforced section.

In 2012, maximum deflections increased on both sections, but the main increment was observed on reinforced section. In 2012, FWD measurement radiuses of the deflection basin located between geophones 1 and 4 have dropped significantly. Furthermore, more pavement deterioration was observed on the evaluated pavement. Although its condition was still satisfactory, the maintenance program has already been initiated. On the one hand, patches and alligator cracks are more frequent on the unreinforced section, but on the other hand, the linear cracks length is higher on the reinforced section. It can be concluded that in this case, positive reinforcing properties of the glass-grid were observed for about 10 years.

The use of a reinforcing grid resulted in an increase in deflection bowl arc section radiuses. Higher radiuses of curvature mean that a greater area of pavement is involved in cooperation under load. It results in lower bending moments in asphalt layers, which translates to a higher pavement fatigue life. It can be assumed that vulnerability to fatigue and bottom-up cracking of pavement was reduced. This positive influence of glass-grids application between asphalt layers is valid, unless the bond between layers is weakened.

The methodology described in this chapter should be repeated for other similar road sections, with different reinforcing materials applied (e.g. made up of mineral fibers and steel nets), having different properties from the glass-grid applied in the tested section. Tests could explain if materials with different parameters lead to the same results as described above.

31.5. Bibliography

[BAŃ 12] BAŃKOWSKI W., GAJEWSKI M., "Accelerated real-scale tests of innovative pavements", *Roads and Bridges no 2/2012*, IBDiM, pp. 89–121, 2012.

[DAV 85] DAVIS T.G., MAMLOUK M.S., "Theoretical response of multilayer pavement systems to dynamic nondestructive testing", *Transportation Research Record*, vol. 1022, pp. 1–7, 1985.

[JEM 08] JEMIOŁO S., GAJEWSKI M., MULARZUK R. *et al.*, "Theoretical models of road meshes and their use in standard pavement designing software", *4th Euroasfalt & Eurobitum Congress*, Copenhagen, pp. 468–475, 2008.

[MAL 09] MALISZEWSKI M., "Assessment of influence of geosynthetic reinforcement to elastic deflection of flexible pavement", *SPENS Seminar– Sustainable Pavements for New Member States*, Warsaw, 2009.

[NCH 08] NCHRP Synthesis Report 381, Falling Weight Deflectometer Usage, 2008.

[SOS 02] SOSN, "Polish system of evaluation of pavement condition", General Directoriate for State Roads and Highways, 2002.

[SYB 11] SYBILSKI D., MALISZEWSKA D., MALISZEWSKI M., "Experiences with reinforcing grids on Warsaw pavement reconstructions", *Construction Materials*, pp. 79–83, 2011.

[ZAW 04] ZAWADZKI J., SKIERCZYŃSKI P., Guidelines for use of geosynthetics in asphalt pavements, Copybook 66/2004 Information, Instructions, 2004.

[ZAW 07] ZAWADZKI J., MALISZEWSKA D., MALISZEWSKI M. *et al.*, "Pavement reinforcement with use of glass-grid", *Drogownictwo*, pp. 329–333, 2007.

32

Effect of Using of Reclaimed Asphalt and/or Lower Temperature Asphalt on the Availability of the Road Network

There is currently a need for a method that assesses the results from changes in the potential durability of road materials due to the inclusion of reclaimed and secondary component materials in the manufacture of new road materials. Such changes will have an effect on the cost of the construction maintenance, both financially to the client and environmentally to society in general, and any savings may be transitory. A site trial has been laid of mixtures with and without reclaimed asphalt and work has started to assess their durability from early-life properties. The trials are being monitored for their initial performance, whereas laboratory trials are concentrating on the combined effect of aging and moisture damage on the performance of asphalt mixtures on the trial. All three strands are being used to develop lifecycle analysis models to customize them for the effect of using alternative component materials on the availability of the network and their overall financial and environmental cost, both initial and whole life. The costs will be identified as being direct (of the construction and maintenance) and indirect (on society in general, such as congestion).

32.1. Introduction

The durability of road materials is an important factor influencing the service lifetime of the road structure or parts of it. Because of its effects on the frequency and extent of maintenance road works, durability plays an important role in

---

Chapter written by Cliff NICHOLLS, Matthew WAYMAN, Konrad MOLLENHAUER, Ciaran MCNALLY, Amir TABAKOVIĆ, Amanda GIBNEY, Aikaterini VARVERI, Sean CASSIDY, Rea SHAHMOHAMMADI and Kevin GILBERT.

the environmental lifecycle performance of the road structure as well as on its lifecycle costs. The EARN project, on which this report is based, was designed to assess the effect that changes in durability of road materials due to the inclusion of reclaimed and secondary component materials in the manufacture of new road materials will have on the cost of the construction, both financially and with regard to the environment.

The project is building upon existing knowledge, supplemented by limited site and laboratory studies, to develop a specific model to look at this issue and to provide indicatory values for use in the model. Existing knowledge has been reviewed to determine the service lifetime of the different pavement layers while a site trial sections mixture with and without reclaimed asphalt has been laid to assess their durability from early-life properties. Laboratory trials are starting, concentrating on the combined effect of aging and moisture damage on the performance of the asphalt mixtures in the site trial. All three strands are being fed into lifecycle analysis models to customize them for the effect of using alternative component materials on the availability of the network and their overall financial and environmental cost.

32.2. Review of existing knowledge

32.2.1. *Relevant parameters for pavement service life*

Asphalt pavement durability is a key factor in determining the performance of a pavement material and, as such, the pavement service lifetime together with the pavement maintenance requirements during its service life. Therefore, it plays an important role regarding the environmental lifecycle of the road structure.

The durability of a pavement involves many relevant parameters that can be categorized as follows:

– the effects from traffic and weather as well as environment and sub-base soil conditions;

– the parameters for unbound base layers, hydraulically bound base layers and bituminous bound base and finally surface layers.

A summary is given in [MOL 13]. However, many data sets are required to evaluate the effect of one parameter on the service lifetime of the pavement. Furthermore, the modeling of a pavement's service lifetime is only possible if most of the parameters are known; otherwise, it is subjected to a wide range of uncertainty.

In pavement management systems (PMS), service lifetimes and qualitative functions for relevant pavement distresses are already incorporated. One approach considers a general design lifetime of the pavement structure of a defined time (e.g. 20 or 30 years) for a design traffic loading. For pavement management, the actual known traffic loading from the beginning of the service life is used for service lifetime prognosis. This approach considers the pavement structure as a whole without utilization in layers and/or materials. Another approach is that some national guidelines already contain service lifetimes for selected pavement materials, which are applied in PMS. In Table 32.1, these assumed service lifetimes derived from various sources are summarized for different pavement materials.

32.2.2. *Service life of low-temperature asphalt mixtures*

Low-temperature asphalt mixtures were developed in order to reduce the paving temperature, the energy used and greenhouse gases emitted. The development of low-temperature asphalt mixtures has been driven by the aim to reduce the temperature effect on asphalt pavement production, laying and compaction, in order to improve the effect of asphalt production on the environment. Furthermore, low-temperature mix asphalts have been utilized to allow the recycling of existing pavements at the end of their service life with reduced demand for material transport, heating energy and raw material consumption.

However, low-temperature asphalt pavements may also result in a shorter service life in comparison to the equivalent standard hot mix asphalt pavements. It is believed that this reduction occurs because the different mixing and paving technologies used in the production of low-temperature asphalt result in weaker mechanical material properties as well as its resistance against cracking.

WMA technologies can be classified in several ways. One way is to classify the technologies by the degree of temperature reduction. Figure 32.1 shows the classification of asphalt mixtures according to the production temperature. Warm asphalt mixes are separated from half-warm asphalt mixtures by the resulting mixture temperature. Specifically, for the warm and half-warm asphalt mixtures, the mixing, laying and compaction are usually undertaken at 100–140°C and at 70–100°C, respectively, whereas for the hot asphalt mixtures the temperatures can reach 138–160°C depending on the bitumen grade used.

Road layer	Pavement material	Germany (FGSV, 2001)		Netherlands (IVON, 2012)		UK (SWEEP, 2014)	
		$\geq$300 ESAL/day	<300 ESAL/day	Right hand lane	Full width	Skid res. lifetime	Structural lifetime
Surface asphalt layers	Asphalt concrete (AC)	12	18	12	18	8	–
	Very thin layer asphalt concrete (BBTM)	–	–	–	–		
	Hot rolled asphalt (HRA)	–	–	–	–		
	Stone mastic asphalt (SMA)	16	22	11	17		
	Mastic asphalt (MA)	19	26	–	–		
	Porous asphalt (PA)	–	–	10	18		
Asphalt base layers	Asphalt concrete (binder layer)	26	30	–	–	–	20
	Asphalt concrete (base layer)	55	75	*	*		
	Other base layers						
	Hydraulically bound base layer	60	80	*	*		
	Unbound base layer	55	75	*	*		
Rigid pavement	Concrete surface layer	26	30	*	*	–	40
	Hydraulically bound base layer	55	70	*	*		
	Asphalt concrete base layer	50	65	*	*		
	Unbound base layer	45	60	*	*		
Maintenance materials	Slurry surfacing	6	8	–	–	8	–
	Micro-surfacing	5	8	–	–		
	Thin hot mix asphalt layer on Sealing	8	10	–	–		

*Highway maintenance in the Netherlands aims at timely strengthening the AC base layers and (sub)bases and, thus, never has to be replaced.

Table 32.1. *General service life assumptions given in guidelines and specifications for pavement management systems*

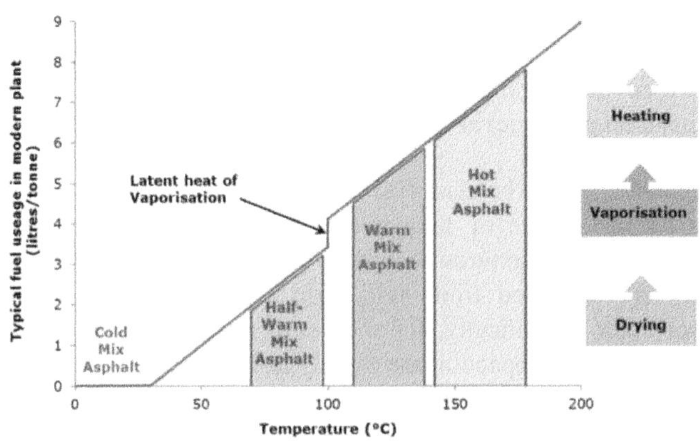

Figure 32.1. *Definition of low-temperature asphalt mixtures [EUR 10]*

A large number of different products that can be used in lower temperature asphalt technologies are currently available in the market. The additives and/or processes used to produce them tend to be proprietary products that may not necessarily have data available that is comparable with that of other products. A list of 35 products that are, or have been, on the market in Europe and/or America has been made [MOL 13].

Despite economic and environmental benefits of the WMA technologies, there are still doubts about its long-term performance [PRO 07]. Prowell *et al.* recommended that further research is needed in order to validate the expected field performance of WMA mixtures, particularity in relation to mix compactibility, rate of gain of structural strength after construction (i.e. curing), rutting, fatigue, moisture sensitivity and the effect of different binder modifiers on the pavement design life.

The potential environmental and social benefits promised by WMA technology will undoubtedly stimulate interest for the wider use of WMA. It may be appropriate to give some advantage to green technologies in the procurement process in order to encourage their use, as were used in the Greenroad project rating system [SOD 07]. However, if the long-term performance of WMA is inferior to HMA, this difference could negate any long-term financial or environmental benefits. Therefore, in its position paper on use of WMA, the European Asphalt Pavement Association stated that WMA procurement should be subjected to lifecycle cost assessment in order to ensure that WMA technologies provide equivalent performance to HMA technologies and that the appropriate maintenance scenarios are fully assessed. Holt [HOL 08] demonstrated that, with a good pavement lifecycle model, significant economic and energy saving can be achieved. He demonstrated that, for highway and national road maintenance cost, reductions can be made with costs reduced by 12% and CO_2 emissions by 56% on a large road maintenance project.

For selected asphalt pavements, the effect of durability in a lifecycle analysis was evaluated during Re-Road project [WAY 12a]. The beneficial effect of applying warm mix technologies can be outnumbered by the effects from extended road maintenance needs if the durability of the pavement is influenced significantly.

32.3. Trial site

32.3.1. *Need for full-size trials*

There are particular advantages that are uniquely associated with full-scale testing over simulated laboratory test programmes [HAR 01]. The effects of size, manufacturing, environment, substructure and loading represent can be better assessed from on-site conditions than from what can be simulated directly with scaled models. However, full-scale tests, which are generally in outdoor ambient conditions, do not permit temperature and moisture to be controlled and, consequently, they are always inherent disparities relating laboratory and on-site test data.

Various pavement monitoring procedures exist [HAR 01], the most basic being the evaluation of structural integrity through monitoring of crack initiation and propagation and permanent deformation on the pavement surface, and more complicated, in-depth evaluations involving monitoring surface and subgrade layers, transverse and longitudinal strains, wheel loads and contact pressures. On-site material properties can be evaluated by obtaining material cores from the road trial and testing it the laboratory for stiffness, water sensitivity aging and fatigue. The falling weight deflectometer can be used to measure on-site surface deflection, and applying appropriate back calculating method stiffness of the pavement layers can be determined.

32.3.2. *Mixture design*

The asphalt mixture investigated in this study was a 10 mm stone mastic asphalt (SMA) typical of that used in Irish and European practice. The variations of the 10 mm SMA mixture are 0% RA as control, 30% RA and no additive, 40% RA and Cecabase RT 945 warm mix additive and 30% RA and Cecabase RT 945 warm mix additive. The grading curves for these mixtures are presented in Figure 32.2, illustrating the good agreement between the control mixture grading and those of the mixtures containing RA. Using the control mixture grading, the guideline allowed the best particle distribution for the mix designs, and consequently the best mixture design as illustrated in Table 32.2.

Mixture Proportional content (%)

no.	RA	10 mm	CRF	Filler	Fresh binder	Warm mix
1	0	65.9	21.8	6.7	5.6	0
2	28.6	43.8	17.0	5.7	4.9	0
3	38.1	34.4	17.1	5.7	4.7	0.5*
4	28.6	43.8	17.0	5.7	4.9	0.5*

*Warm mix additive added to mixtures 3 and 4 at 0.5% of the total binder content in the mixture. CRF, crushed rock fines.

Table 32.2. *Mixture designs*

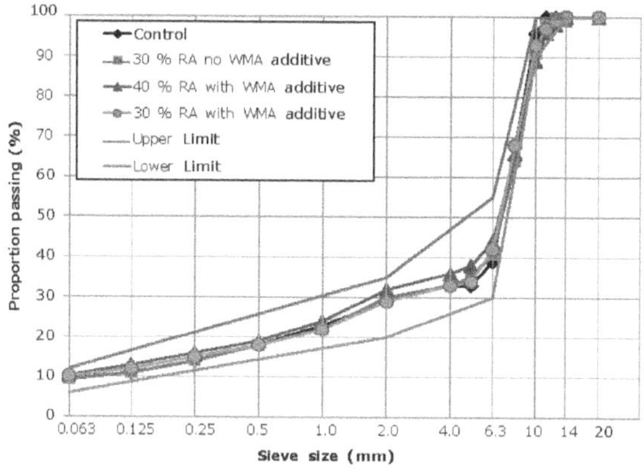

Figure 32.2. *Particle size distribution. For a color version of this figure, see www.iste.co.uk/jacob/materials.zip*

32.3.3. *Selection and construction of the test section*

In collaboration with the Irish National Roads Authority, a section of the N3 national road was identified as a suitable road section for the site trial experiment. The site was located between Blanchardstown and Clonee village, at the outskirts of the Dublin city. The GPS coordinates of the trial site are latitude 53° 24′ 19.35″, longitude –6° 24′ 30.55″ to latitude = 53° 24′ 6.43″, longitude = –6° 23′ 59.21″. The section was chosen because the road section was due for resurfacing, it is close to the asphalt plant (ca. 60 km) and it is on a main commuter route into Dublin city with an average daily vehicle traffic count, one direction only, of 15,480 vehicles including HGV (http://nraextra.nra.ie/CurrentTrafficCounterData/index.html). Figure 32.3 illustrates a satellite image of the trial section and surrounding area. The road is a dual carriage way with three traffic lanes on each side (bus lane and two traffic lanes).The middle lane was chosen as the test lane because it will be subjected to the most trafficking, particularly from heavy goods vehicles. The traffic direction is toward Dublin city. Figure 32.4 shows a schematic layout of the trial section. The site was split into four sections of varying lengths for the different mixtures.

Figure 32.3. *Satellite image of the trial road section*

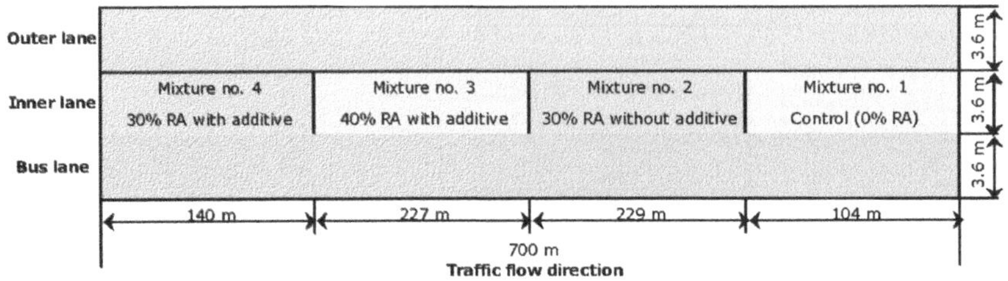

Figure 32.4. *Schematic representation of the trail section*

To cover the trial section area, it was estimated that just over 230 tonnes of asphalt material was required. The work started with the removal of the existing surface course, which was milled to a depth of 40 mm. An initial regulating course was then laid to a depth of 20 mm. The outer lane and bus lane (Figure 32.4) were resurfaced with a standard SMA, containing no RA or warm mix additive, to a depth of 40 mm. The test lane was resurfaced with the materials described above.

The paving process started with laying section 1 (control mixture). The asphalt material was hauled from the plant to the site by truck and unloaded to the material transfer vehicle before it was sent to the paver. The purpose of the material transfer vehicle was to remix the material before sending it to the paver and laying it onto the road. Figure 32.5 shows the paving process. The paving process of section 1, passed as expected without any difficulties. However, section 2 proved to be more difficult because the mixture was cooling down rapidly with the consequential reduction in workability of the mixture. The paving of sections 3 and 4 passed without much difficulty, highlighting the improved workability of the mixtures incorporating the warm mix additive, with up to 40% RA. The site work records are summarized in Table 32.3, giving section lengths, temperature and weight of each mixture.

32.3.4. *Reclaimed asphalt feedback*

The reclaimed asphalt feedstock was supplied from a site on the M1 motorway in North County Dublin and is derived from a single source 14 mm porous asphalt. The material was milled and stored in a depot until required on this project. The total amount of reclaimed asphalt material supplied was 170 tonnes. The quantity of the processed reclaimed asphalt material by size is given in Table 32.4. The visual inspection revealed that the >16 mm material contained binder course material aggregate. Therefore, the >16 mm and <6 mm reclaimed asphalt aggregate was screened out and not used for the trial asphalt mixtures.

Figure 32.5. *Paving process of the trial section. For a color version of this figure, see www.iste.co.uk/jacob/materials.zip*

Mix no.	RA content (%)	Containing warm mix additive	Load no.	Start chainage (m)	End chainage (m)	Discharge temperature (°C)	Rolling temperature (°C)	Weight (tonnes)
1	0	No	1	0	104	150	134	30.00
2	30	No	2	104	155	115	105	17.20
			3	155	220	130	115	17.20
			4	220	333	150	130	28.90
3	40	Yes	5	333	385	137	125	30.10
			6	385	458	135	125	17.00
			7	458	560	134	128	28.80
4	30	Yes	8	560	618	125	118	17.00
			9	618	672	132	124	17.20
			10	672	700	136	128	28.65

Table 32.3. *On-site work record of asphalt material*

Size (mm)	>16	16–12.5	12.5–6	<6
Quantity (T)	40	45	35	50
Proportion of total (%)	24	26	21	29

Table 32.4. *The quantity of the processed RA material by size*

The binder content in the RA was determined according to the EN 12697-39. Five samples of RA were taken and weighed. The samples were placed in the oven at 530°C for 30 min. Once the samples were cooled, they were weighed again and proportion of binder in the mixture was calculated. The average binder content was 5.3%. Following the binder burn off procedure, the material particle size distribution determined following EN 12697-2. The reclaimed asphalt material aggregate size distribution/grading is shown in Figure 32.6 and the binder contents were 5.2, 5.4, 4.8, 5.7 and 5.4% with an average of 5.3%.

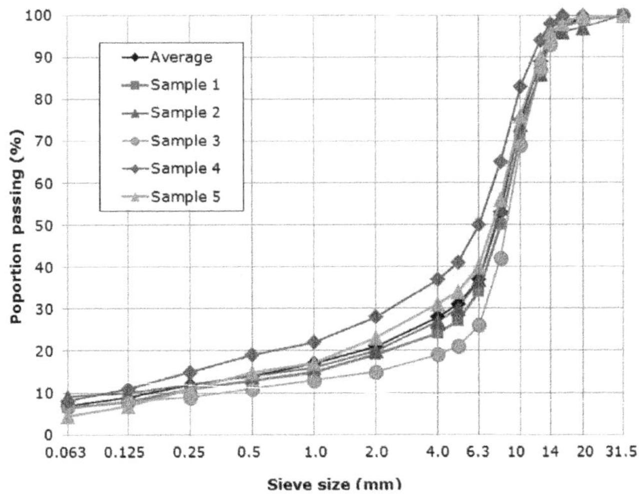

Figure 32.6. *RA material grading after the binder removal*

32.3.5. *Testing*

The asphalt mixture material was sampled from the paver as shown in Figure 32.7(a); this material is used for stiffness and water sensitivity testing at a later stage. In addition, a total of 108 cores (27 from each trial section) were taken 24 h after the construction was completed. These cores will be used for subsequent laboratory moisture conditioning by the moisture-induced sensitivity tester (MIST) device. The coring procedure is shown in Figures 32.7(b) and 32.7(c). The laboratory testing procedures are further described in section 32.4 of this chapter.

As part of the study, the ride quality of the test site will be investigated. A number of tests will be conducted to allow evaluation of the international roughness index (IRI), pavement conditioning index and skid resistance. At the time of writing, only the IRI testing had been completed. Continuous measurements of IRI were acquired using the road surface profiler. The data were collected at speeds of

70–80 km/h in order to ensure that there was no delay or disruption for other road users. The entire data collection process was non-contact using high-frequency lasers and accelerometers in conjunction with a very accurate distance measurement system. The IRI measurements were continuously measured in both inner and outer wheel paths and the values were recorded at 10 m intervals. Table 32.5 shows average IRI values for the test sections. The average IRI value for each section is < 2, which shows good ride quality of the pavement surface.

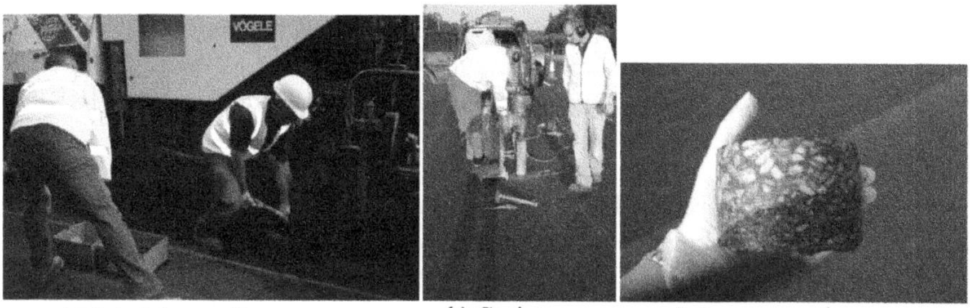

a) Material sampling b) Coring c) A core sample

Figure 32.7. *Collection of samples from site*

Section	IRI (m/km)		
	Left	Right	Average
1	1.25	1.23	1.24
2	0.96	0.99	0.98
3	1.04	1.10	1.07
4	1.33	1.43	1.38

Table 32.5. *Average IRI values of the test site*

32.4. Laboratory testing

During their service life, the performance of asphalt pavements is greatly affected by different environmental factors such as moisture, oxygen, heat, pressure and UV light. Primarily in this work, moisture damage and aging due to oxygen diffusion were assumed to be the most important parameters that can shorten the pavement life and accelerate pavement distresses. The use of reclaimed asphalt, secondary aggregates and lower temperatures will affect the various factors in different ways, further complicating an already complex situation. Presently, laboratory trials are starting, concentrating on the combined effect of aging and moisture damage on the performance of the asphalt mixtures in the site trial.

Another important parameter that is considered to be a serious cause for diminishing the long-term performance of asphalt concrete pavements is moisture damage. Regardless of the mix composition, asphalt mixtures with moisture will suffer from moisture damage at some time during their service life. At the micromechanical level, moisture-induced damage is described as the loss of strength and durability of the asphalt mixture through loss of adhesion between the asphalt binder and the aggregates, as well as through loss of cohesive resistance of the asphalt binder itself [KIG 98]. Moisture diffusion, desorption of the asphalt binder due to fast water flows and cyclic pore pressure development from entrapped water in the air voids (i.e. pumping action) were identified as the physical and/or mechanical processes that ultimately can lead to pavement distresses [KRI 05, KRI 08].

In this project, moisture diffusion and pumping action were identified as the dominant moisture damage inducing mechanisms. However, the time frame over which each mechanism occurs in the field differs significantly. Moisture infiltration occurs over a longer timeframe, while excess pore pressure development takes place in very short times. In order to address the individual damage mechanisms associated with these two moisture damage inducing processes, the test protocol utilized consists of a combination of two different conditioning methods: (1) bath conditioning and (2) cyclic water pore pressure application. Bath conditioning is performed at elevated temperatures so as to facilitate the infiltration of water into the asphaltic mixture and, consequently, to accelerate the long-term degradation of the material properties. Cyclic pore pressure generation in the asphalt mixture is achieved by means of the MIST. The MIST was designed as an accelerated conditioning device for the evaluation of the resistance of an asphalt mixture to stripping by simulating the high pressure fields, which develop within an asphalt layer due to traffic loading [INS 09].

In order to evaluate the material used on-site, a series of laboratory tests are planned, including the indirect tensile stiffness modulus (ITSM) test to EN 12697-26, a water sensitivity test to EN 12697-12 and the MIST. The ITSM and water sensitivity test will be carried out at four stages starting from the initial trial construction, with further testing scheduled for 3, 6 and 12 months. The initial testing stage is underway at the time of writing using material collected from the paver during the construction process. These are later compacted in the laboratory using the gyratory compaction procedure given by the EN 12697-31.

For MIST testing, cylindrical samples of 100 mm in diameter and 50 mm in thickness cored out from the site trials are obtained. The specimens first will be subjected to moisture infiltration by placing them in a bath filled with distilled water at a temperature of 60°C. At fixed time intervals of 0, 3 and 6 weeks, three specimens per mixture are removed from the bath, placed in a bath at 20°C for 2 h and then stored at a climatic chamber at 20°C until tested for their strength using the indirect tension test (ITT). The rest of the samples are further conditioned in the MIST device by applying 3,500 cycles of water pumping action at a temperature of 60°C and a pressure of 0.48 MPa. After MIST conditioning, the samples are placed into a water bath at 20° C and kept there for 2 h before ITT testing.

32.5. Assessment of lifecycle cost and carbon footprint

32.5.1. *Assessment on a lifecycle basis*

Lifecycle cost and contribution to climate change of the mixtures will both be assessed on a lifecycle basis. The applicable standards will be followed for each assessment; ISO 14067:2013 for the carbon footprint and ISO 15686-5 for the lifecycle costing. In addition, calculation of the carbon footprint will be facilitated using an updated version of the asphalt pavement embodied carbon tool [WAY 12b], a tool that meets the appropriate standards and is specific to asphalt life cycle, providing the necessary level of granularity to assess the impacts of the inclusion of recycled content, lower temperature production and variable durability. Cost and carbon impacts will be assessed over a 60 year analysis period, allowing for the effects of multiple asphalt renewal cycles to be observed. The rates of renewal will be directly informed by the laboratory assessment outlined in section 32.4 and early observations of the trial site. Furthermore, both assessments will consider multiple repetitions of raw material acquisition to product production, installation, maintenance and end of life.

32.5.2. *Formulation of the carbon footprint*

Although the lifecycle cost assessment can be informed by data that are readily available to plant and infrastructure managers, some of the data to inform the carbon footprinting assessment need to be collected at first hand, while the asphalt production takes place. To this end, minute-by-minute readings of energy consumption at the Lagan Kinnegad production plant were taken during production of the four mixtures detailed in Table 32.3 in order to truly observe the consequences of lower temperature mixing with recycled content during a typical batch plant production cycle. Having observed and collected data from a typical batch production cycle, where the plant is heated only once from ambient temperature to produce all four mixtures consecutively, there is a need to apportion the overall energy consumption (both gas oil and electricity) equitably between the mixtures according to their particular characteristics, including the recycled asphalt content and heating temperature, having first accounted for factors common to all mixtures, such as the efficiency of the plant and the associated heat loss. The outcome of this analysis will be a methodology that can be applied to future trial mixtures produced in a similar manner, a quick and easy method to investigate the carbon footprints of trial low temperature asphalts.

Plant energy consumptions will be one component of the cradle-to-site carbon footprints of the materials and these will be combined with the contributions of raw materials, transport and installation works. Later in the project, the durability of the materials will be factored in order to extend the analyses from cradle to end of life.

32.6. Conclusions

This chapter describes the initial work of a project to develop a model that can be used to assess the effect that changes in durability of road materials due to the inclusion of reclaimed and secondary component materials in the manufacture of new road materials will have on the cost of the construction, both financially and with regard to the environment. The work has not been completed at the time of writing, but the data necessary to build the model is being collected.

32.7. Acknowledgments

The work reported here was part of a project funded by the CEDR Transnational Road Research Programme Call 2012: Recycling: Road construction in a post-fossil fuel society that is being funded by Denmark, Finland, Germany, Ireland, the Netherlands and Norway and managed by Nation Road Authority, Ireland. The project started in January 2013 and was completed by December 2014.

32.8. Bibliography

[EUR 10] EUROPEAN ASPHALT PAVEMENT ASSOCIATION, "The use of warm mix asphalt. EAPA position paper", EAPA, Brussels, available at http://eapa.org/usr_img/position_paper/the_use_of_warm_mix_asphalt_january_2010.pdf, 2010.

[OBJ 12] OBJECTBEHEERREGIME VERHARDINGEN, IVON, "Fiola van der, P. Version 1.1", Rijkswaterstaat, Delft, 2012.

[FOR 01] FORSCHUNGSGESELLSCHAFT FÜR STRAßEN UND VERKEHRSWESEN, *Richtlinie für die planung von erhaltungsmaßnahmen (Guideline for the Planning of Conservation Measures)*, FGSV, Köln, 2001.

[HAR 01] HARTMAN A.M., GILCHRIST M.D., OWENDE P.M.O. *et al.*, "In-situ accelerated testing of bituminous mixtures", *Road Materials and Pavement Design*, vol. 2, no. 4, pp. 337–357, 2001.

[HOL 08] HOLT C., "New and better tools for life-cycle assessment", *Proceedings of Sustainable Transport Forum Workshop*, Toronto, Canada, May 28–29, 2008.

[INS 09] INSTROTEK, "Moisture induced sensitivity test operators guide, Version 1", Raleigh, NC, 2009.

[KIG 98] KIGGUNDU B.M., ROBERTS F.L., "Stripping in HMA mixtures: state-of-the-art and critical review of test methods", NCAT Report 88-02, National Center for Asphalt Technology Alabama, Auburn University, 1998.

[KRI 05] KRINGOS N., SCARPAS A., "Raveling of asphaltic mixes due to water damage: Computational identification of controlling parameters", *Transportation Research Record Journal of the Transportation Research Board*, vol. 1929, pp. 79–87, 2005.

[KRI 08] KRINGOS N., SCARPAS A., KASBERGEN C. *et al.*, "Modelling of combined physical mechanical moisture-induced damage in asphaltic mixes, Part 1: Governing processes and formulations", *International Journal of Pavement Engineering*, vol. 9, no. 2, pp. 115–118, 2008.

[MOL 13] MOLLENHAUER K., NICHOLLS J.C., VARVERI A. *et al.*, *Service Lifetime, Suitable Test Methods for Characterising and Main Parameters Controlling Durability of Warm-Mix Asphalt Containing RA*, TRL, Wokingham, available at http://www.trl.co.uk/EARN/, 2013.

[PRO 07] PROWELL B.D., HURLEY G.C., CREWS E., "Field performance of warm mix asphalt at the NCAT track test", *86th Annual Meeting of Transportation Record Board*, Washington, DC, 2007.

[SOD 08] SODERLUND M., MUENCH S.T., WILLOUGHBY K. *et al.* "Green roads: a sustainability rating system for roadways", *87th Annual Meeting of Transportation Record Board*, Washington, DC, 2008.

[WAY 12a] WAYMAN M., PARRY T., ANDERSSON-SKÖLD Y. *et al.*, "Lifecycle assessment of reclaimed asphalt", Re-Road deliverable D3.4, available at http://re-road.fehrl.org, 2012.

[WAY 12b] WAYMAN M., SCHIAVI M.I., CORDELL B., *et al.*, "The asphalt pavement embodied carbon tool (asPECT): developing a carbon footprinting methodology for asphalt products", *5th Eurasphalt & Eurobitume Congress*, Istanbul, Turkey, 2012.

33

Brazilian Road Deterioration Test: Final Report

The practice of overloading increases as the control of heavy vehicle decreases. The purpose of this study is to do a continuous road assessment concerning strains, stress, moisture and temperature of pavement as well as recording dynamic load effort. This work is a result of a study developed in an experimental track located near Araranguá city, in the south of Brazil. The first activity considers pavement instrumentation, with sensors installed in each pavement layer and weigh-in-motion sensors installed on the pavement surface. The second activity consists of performing the fatigue test and the complex modulus determination. The pavement mechanical behavior is evaluated in the laboratory, which provides information to the pavement deterioration model. Then, with traffic load spectra, analysis for the pavement life will be performed. The load spectra values are transformed into a pavement deformation response to be compared to the pavement fatigue curve values. The pavement deterioration model is then loaded with different loads characteristics to simulate different conditions of deterioration.

33.1. Introduction

The development of a high-speed multiple sensor weigh-in-motion (WIM) system has been researched by several international organisms and countries. For Brazil, as for most countries, WIM systems are an important tool to perform weight enforcement directly on the highway. This project is inserted in a broader study called the high-speed multiple sensor WIM systems and pavement mechanical analysis. This project came into existence due to the National Department of Infrastructure and Transportation (DNIT) needing to perform weight control in the National Weighing Plan. Then, an agreement between DNIT and

---

Chapter written by Gustavo Garcia OTTO, Leto MOMM and Amir Mattar VALENTE.

Federal University of Santa Catarina (UFSC) was established to elaborate and define MS/WIM procedures. In Brazil, these new technologies have the potential to perform statistics data collection, overload pre-selection and, in the future, direct enforcement. One of the most important contribution for WIM development and use as reference for this research is the COST 323 research project [COS 98].

The WIM systems allow us to estimate the resultant force on each vehicle axle. Additionally, it has great potential to increase the efficiency and effectiveness of control overloading practices, which are applied by the authorities responsible for road operations. This work aims to study pavement deterioration in real conditions by carrying out pavement assessments, using both information from pavement instrumentation and multiple sensor WIM systems to define a pavement deterioration model.

The road mechanical behavior study provides a better understanding of the physical phenomena involved in the axle force. The pavement structural response, as vehicles travels on the pavement surface, is defined by stress and strain efforts at a given temperature and moisture conditions. The activities applied to build the mechanical behavior concept are performed both in the field and laboratory.

In field, two independent data acquisition systems provide strain and stress information. The recorded data correspond to the pavement structural layers behaviors as well as the dynamic force from the free flow traffic. In the laboratory, materials from the test site allow to determine mechanical characteristics of material, mainly by the fatigue test and complex modulus determination, both performed in the Pavement Laboratory at UFSC.

The sensors are making reference to both the average wheel tracks shown on the road segment chosen. Sensors installed on these two data acquisition points have different depths corresponding to the layers interfaces. The first group is constituted by pairs of strain gauges in longitudinal and transversal directions. The second group is constituted by pressure cells, placed on each pavement layer. In the third group, a group of moisture sensors are installed at three different depths. It uses dielectric permittivity that corresponds to the water volume content inside the granular layers. One WIM sensor, made with optical technology, is used to collect axle weight information for better understanding of the pavement mechanical behavior. Finally, a group of three temperature sensors are installed inside the asphalt concrete.

At the end of the project, a mathematical model is constructed using numerical analysis software, which allows running simulation analysis according to conditions found in the field. The overload practices are considered to be the major cause of pavement deterioration. The results presented here show the steps for building a model of deterioration of a road segment.

33.2. Field analysis

In field, the main objective is to analyze a pavement structure in real conditions of load and climate. The results will provide a better understanding of the complexity of field conditions. For this project, a 700 m long experimental track was built, located at BR-101 highway (418 km), near Araranguá, Santa Catarina, Brazil. The cross-section is composed of a semi-infinite sand subgrade, a 20 cm Macadam layer, an 18 cm graded gravel layer and a 17 cm Hot Mix Asphalt layer. It was designed to be the weigh station main entrance, as shown in Figure 33.1.

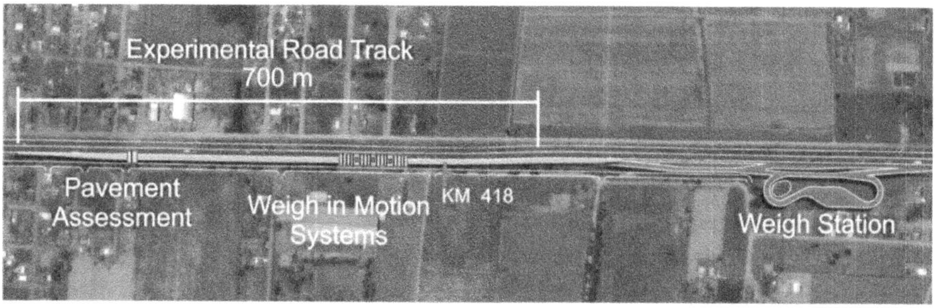

Figure 33.1. *Location of test track, pavement assessment and weighing in motion systems*

The road track was instrumented with sensors located at two different places. The first group of sensors is related to the pavement assessment, installed in February 2011. The second group corresponds to the WIM sensors, installed in April 2009. The information from both systems complete each other, thus creating a deterioration profile. It allows for correlating the axle dynamic force with pavement structure response.

The pavement structure fatigue process has its origins in the repeated interactions between the tire and the pavement. The pavement material behavior, allied with external factors as solicitation and physical conditions, allows us to understand the mechanical process of pavement deterioration. Pavement strain and stress assessment shows the intensity of the efforts applied. It can be used to evaluate the pavement support parameters. Pavement design aims to obtain an appropriate behavior during lifetime, as well as ensure safety and comfort of the road users. This process considers the load applied over time, the deformation found in the bottom of surface course and the stress found in the top of subgrade.

The pavement assessment system uses devices under and over the pavement surface, which allow for the acquisition of information for the identification and quantification of pavement behavior under certain conditions of temperature and humidity. A set of sensors are installed in a way that allows for identifying the deformation under asphalt layers, base layers and sub-base , in longitudinal and transverse alignments. Likewise, pressure on top of the base layer, sub-base and subgrade. Moisture of granular layers and temperature at three depths in the asphalt concrete are measured at same time as the other measures. The longitudinal and transversal strains, in the bottom of the layers, are monitored by strain gauge specially designed for this function.

The vehicle weighing data derives from the WIM system. The WIM sensors are installed in rows in transversal aliment to the direction of the traffic. The technologies used are piezoelectric quartz, piezoelectric ceramic, piezoelectric polymer and fiber optic sensors.

Weighing systems provide information about the fleet on the experimental track. The sensors are installed in groups of 16 rows spaced at 1 m between each line. The data of the weighing system are organized by date and time, which is recorded when a vehicle met the first sensor. Each event contains information of the vehicle class, axle numbers and space between them, axle weight, gross weight and speed.

33.3. Pavement deterioration analysis

The analysis of the deterioration of pavements uses both measurements of the pavement assessment system and the WIM system together with material characteristics defined in the laboratory. The sensors installed in the pavement structure informs the pavement behavior in the moment that one vehicle passes by. As vehicles pass over the experimental track, the WIM system identifies and measures their axle weight.

A numerical model built with pavement material characteristics was determined in the laboratory and field. The material characteristics defined in the laboratory are complex modulus and fatigue curve. These two characteristics govern structural behavior of the pavement during the lifetime span. The pavement structure is assessed applying both traffic information from WIM system data collection and data traffic from the weigh station.

33.3.1. *Asphalt mechanical characteristics*

Asphalt mechanical characteristics are found with an alternated bending machine. This equipment allows for determining the complex modulus and the fatigue characteristics in different conditions of temperature and load frequency. The results of complex modulus determination give the linear viscoelastic characteristics of the asphalt material, because they are measured in the field of small deformations. All viscoelastic materials have a response delay related to the tension applied. In the field of small deformations, the small sinusoidal force results also in a sinusoidal response. Thus, the deformation presents a phase shift in relation to the force, which reflects the material behavior. When the phase shift is equal to 0°, it means that the material is purely elastic. When phase shift is equal to 90°, it means that the material is purely viscous. The modulus of the complex $|E^*|$ can be express by equation [33.1]:

$$|E^*| = (E_1^2 + E_2^2)^{1/2} \quad [33.1]$$

The different components of the complex modulus vary with temperature and frequency. The experimental results $|E^*|$, φ, E_1 and E_2 are expressed by graphical representations called isotherme, equivalent frequency, isochrone, cole–cole and black space. E_1 represents the real part, or elastic part, and E_2 represents the imaginary part, or viscoelastic part. The specimen is a block with dimensions (h_{layer}) 17 cm × 60 cm × 40 cm removed from the asphalt course, and by using this trapezoidal specimens are made to perform complex modulus and fatigue tests in the two-point bending machine.

The cole–cole representation provides the parameters for the Huet–Sayegh model [HUE 63], which provides parameter for pavement designand calculated using Viscoanalyse (Table 33.1, IFSTTAR, 2009). The model is obtained by a spring of stiffness E_0, which represents the elastic modulus. The model is given by equation [33.2]:

$$E^*(i\omega\tau) = (E_0 + (E_\infty - E_0))/(1+\delta(i\omega\tau)^{-k} + (i\omega\tau)^{-h}) \quad [33.2]$$

E_0	E_{inf}	Delta	K	h	τ	C1	C2	T.ref	A0	A1	A2
1,254	13,139	0.573	0.132	0.518	0.065	21,433	195,397	15	1,405	–0.295	0.0013

Table 33.1. *Huet–Sayegh parameters*

Figure 33.2 shows complex modulus representation of the asphalt mixture from the experimental road track.

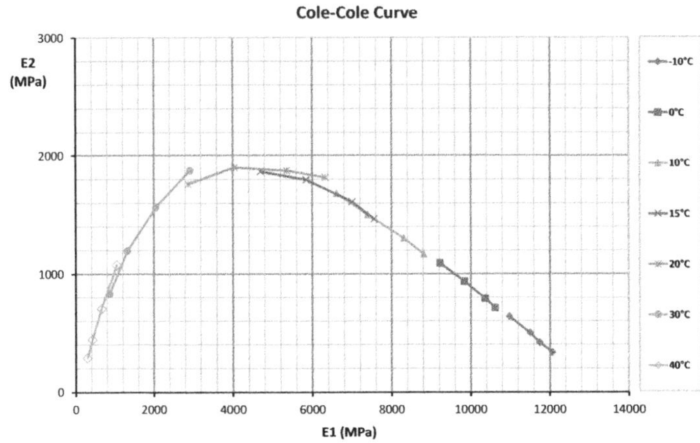

Figure 33.2. *Cole–cole complex modulus representation. For a color version of this figure, see www.iste.co.uk/jacob/materials.zip*

The fatigue test is performed in continuous mode and controlled deformation, thus the stiffness of the specimen decreases as the number of deformations applied increases. The results must be around the 10^6 numbers of alternated deformations in the two-point bending test. The test aims to find a deformation corresponding to one million cycles applied, called ε_6. The criteria adopted are the same as recommended by the French fatigue test method, executed at 10°C and 25 Hz.

The fatigue curve is presented in Figure 33.3. The curve represents the susceptibility of asphalt mixture to deformation effort, represented in scale log–log. The abscissa is the deformation, in 10^{-6} m, and the ordinate is the accumulated number of deformation cycles. The curve is a line represented by an exponential equation, in which curve inclination is the exponent. The fatigue equation can be written as:

$$N = 1.22 \times 10^{16} \times \text{Def}^{-5.10} \quad [33.3]$$

The deformation-specific result of the fatigue test result is the deformation for 10^6 cycles. When the number of cycle N is substituted in equation [33.3], it gives the deformation ε_6 equals to 95.024 μm. N represents the number of axels accumulated on the pavement over the years; in other words, the sum of loads, which will pass by the road section during the pavement lifetime. Therefore, the equation enables comparing the deformation found in field with the fatigue curve to establish the deterioration degree.

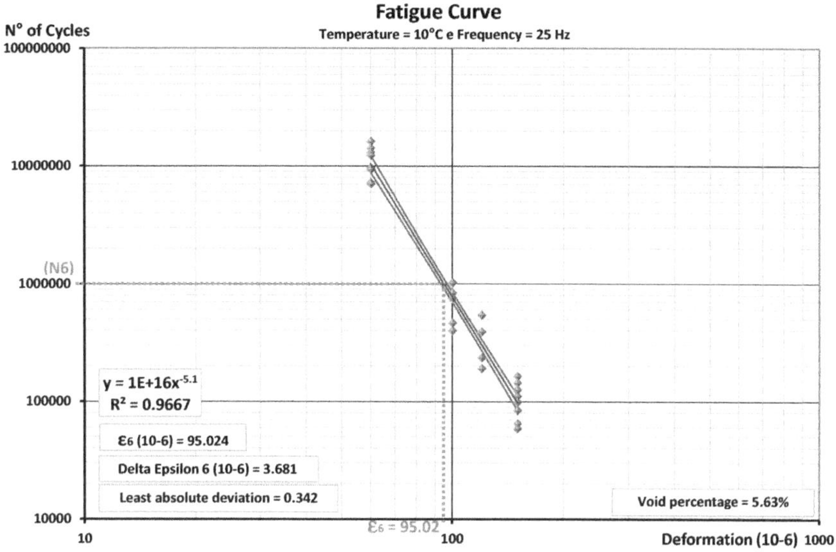

Figure 33.3. *Asphalt fatigue curve. For a color version of this figure, see www.iste.co.uk/jacob/materials.zip*

33.4. Analysis model

The analysis model considers that there is a traffic load spectra (load per axle or axle group) for each load, speed and pavement temperature. A computer program calculates the equivalent strain spectra, which is dependent on the load speed and temperature. The strain frequency spectrum shows the distribution of deformation applied on the pavement due to passing traffic combined to pavement characteristics (a composition of load and speed).

The calculus of the damage is a function of the load (*C*), velocity (*V*) and temperature (*T*) as shown in equation [33.4]. The damage is cumulative sum of damages over the life of the pavement that breaks when the value *D* is equal to 1.

$$D = \sum_{k}^{C} \sum_{i}^{V} \sum_{j}^{T} \frac{n_{ijk}}{N_{ijk}} \times 100 \qquad [33.4]$$

In the equation, *n* is the cumulative traffic load summary over pavement lifetime and *N* is the number of load accumulated that leads to end of pavement life according to a certain amount of deformations. The fatigue curve can connect the pavement lifetime, in this analysis in years, with the amount of load applied.

33.5. Results

The heavy vehicle fleet represents 83% of total volume and it consists of five main vehicle classes presented in Table 33.2.

Vehicle class (according to the Brazilian legislation)	Description	Composition (%)
3C	Three-axle vehicle:onesingle and one tandem (single unit)	26.34
3S3	Six-axle vehicle: one single, one tandem andone tridem	15.48
2S3	Five-axle vehicle:one single, one dual tire andone tridem	15.23
2C	Two-axle vehicle	14.48
2S2	Four-axle vehicle:one single, one dual tire andone tandem	11.24

Table 33.2. *Vehicle class composition*

Four types of axes stand out from all existing axes: single axle single wheel, single axle dual tire, tandem axle and tridem axle. Vehicle fleet has its specific axle composition and, for each axle type, there is a respective load distribution. The loads in field obey a normal distribution, with the most common value near legal limit and other values below or above the expected value. Figure 33.4 shows the load spectra for the three most influent types of axles and their legal limits (described in the Brazilian legislation).

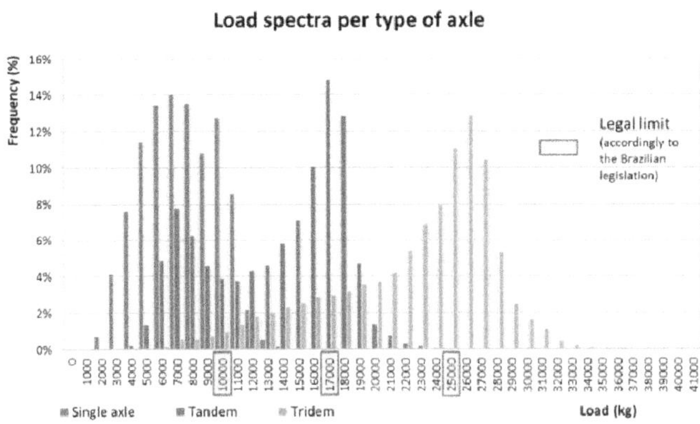

Figure 33.4. *Axle load spectra*

All parameters are organized and loaded into the analysis software together with pavement structural characteristics. Among all results, the pavement longitudinal deformation is the behavior observed. Then, this value can be linked directly to the deformation found in the fatigue curve. The design information used for simulation of the traffic and the behavior of the pavement are as follows:

- traffic open year 2011;
- project period of 15 years, 2026;
- characteristics of asphalt concrete (Model rheological Huet–Sayegh);
- characteristics of the base BG, sub-base and subgrade MS;
- axle types: single axle dual tire, tandem and tridem;
- actual average temperature of the pavement ($z = 17$ cm): 22°C;
- speed: 80 and 60 km/h.

Table 33.3 contains the physical and statistical parameters used in the pavement numerical simulation. The coefficients K_r, K_c and K_s correct the deformation ε_6 from the fatigue test and then become working strain at the base of the bituminous layer.

	Parameters	Value
Complex modulus	$\|E^*\|$ (10°C,10 Hz)	8,416
	$\|E^*\|$ (10°C, 25 Hz)	8,440
Fatigue	$\|E^*\|$ (15°C, 10 Hz)	7,186
	ε_6 (10^6)	95.024
Coefficient	Fatigue slope	–5.1
	Kr	1.0
	Kc	1.1
	Ks	1.0

Table 33.3. *Physical and statistical parameters*

33.5.1. *Axle interaction simulation*

Two simulations can present how pavement life is related to load and velocity variations. The analysis uses the correspondent strain considering a fleet of 100% of single axel, 100% of tandem axle and 100% of tridem axle. Table 33.4 presents the result from traffic and pavement life interaction, considering increasing the load, starting from legal limit, and vehicle velocity variation.

Axle	Load (t)	εt (obtained) (10^{-6}) 80 km/h	εt (obtained) (10^{-6}) 60 km/h	End of life 80 km/h	End of life 60 km/h
Single axle	8.2	49.9	51.0	2,026	2,023
	10	60.8	62.1	2,013	2,013
	10.5 (5%)	63.8	65.2	2,011	2,011
	10.75 (7.5%)	65.4	66.8	2,011	2,011
Tandem axle	17	49.7	50.9	2,026	2,023
	17.85 (5%)	52.2	53.5	2,022	2,020
	19.19 (7.5%)	56.1	57.5	2,017	2,016
Tridem axle	25	48.3	49.6	2,026	2,023
	26.25 (5%)	50.8	52.0	2,023	2,023
	28.22 (7.5%)	54.6	55.9	2,019	2,017

Table 33.4. *Pavement end of life proportional to load increase and velocity variation*

The study found that for any increase on the load above the legal limit decreases pavement life. In addition, velocity variation has an important effect on the damage caused on pavement structure; the lower velocities, the bigger the damage.

An analysis considering the wide range of velocity and load shows how important the velocity is to pavement life. The analysis resulted for each type of axle is given in the following tables. Table 33.5 shows the pavement end of life due to 100% fleet of vehicles with single axle.

Load increase (%)	40 km/h	60 km/h	80 km/h	100 km/h	120 km/h
0	2,013	2,013	2,013	2,014	2,014
5	2,012	2,012	2,013	2,013	2,013
7.5	2,012	2,012	2,012	2,012	2,013
10	2,011	2,012	2,012	2,012	2,012
12.5	2,011	2,011	2,012	2,012	2,012
15	2,011	2,011	2,011	2,012	2,012

Table 33.5. *Pavement end of life proportional to load increase and velocity variation*

Table 33.6 shows the pavement end of life due to 100% fleet of vehicles with tandem axle.

Load increase (%)	40 km/h	60 km/h	80 km/h	100 km/h	120 km/h
0	2,018	2,019	2,020	2,020	2,022
5	2,016	2,017	2,018	2,018	2,020
7.5	2,015	2,016	2,017	2,017	2,019
10	2,015	2,016	2,016	2,017	2,018
12.5	2,014	2,015	2,015	2,016	2,017
15	2,014	2,014	2,015	2,015	2,016

Table 33.6. *Pavement end of life proportional to load increase and velocity variation*

Table 33.7 shows the pavement end of life due to 100% fleet of vehicles with tridem axle.

Load increase (%)	40 km/h	60 km/h	80 km/h	100 km/h	120 km/h
0	2,019	2,020	2,021	2,022	2,022
5	2,017	2,018	2,019	2,020	2,020
7.5	2,016	2,017	2,018	2,019	2,019
10	2,015	2,016	2,017	2,018	2,018
12.5	2,015	2,016	2,016	2,017	2,017
15	2,014	2,015	2,016	2,016	2,016

Table 33.7. *Pavement end of life proportional to load increase and velocity variation*

33.5.2. *Real traffic composition interaction*

Consider an entire fleet of vehicle composed of 100% of traffic with a specific load is too radical, but indicates how important an overload of a specific axle is comparing to others. As we can see in Figure 33.5, even when we consider a real traffic data and an increase of 5, 7.5, 10 and 20% of overload.

If all vehicles start to carry more than 5% of overload, the pavement life reduces to 15%; if they carry more than 7.5% of overload, the pavement life reduces to 22%; and if they carry more than 10 and 20% of overload, the pavement life reduces to 28 and 50%, respectively.

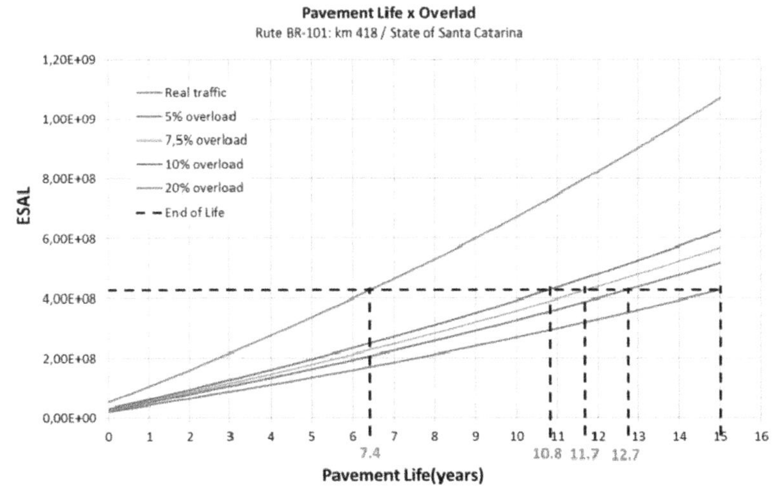

Figure 33.5. *Pavement life × overload. For a color version of this figure, see www.iste.co.uk/jacob/materials.zip*

33.6. Conclusions

It can be assured that any increase on the load over the axles is the major problem for pavement performance and consequently contributes to reduce pavement life. These phenomena can be observed in the first analysis. According to the material viscoelastic behavior law, faster the load is applied harder the material behaves. As shown in the results, lower is the vehicle speed shorter the pavement lifespan is. The real traffic composition shows the increase in overload and pavement life reduction; it is clear that 20% of overload can reduce pavement life in approximately 50% of cases.

33.7. Acknowledgments

We would like to thank all team members, who worked very hard to install, manage and control data collection at the experimental site.

33.8. Bibliography

[CEB 93] CEBON D., *Interaction Between Heavy Vehicles and Roads*, L. Ray Buckendale Lecture, SAE trans 93001. SP-951, Cambridge University Engineering Department, Cambridge, UK, 1993.

[COS 98] COST 323, "Weigh-in-motion of road vehicles: final report", Institut Francais des Sciences et Technologies des Transports, de l'Amenagement et des Reseaux (IFSTTAR). Paris, France, 1998.

[DE 96] DE LA ROCHE C., Modulus de rigidité et comportement en fatigue des enrobés bitumineux: Expérimentations et nouvelles perspectives d'analyse, PhD Thesis, Ecole Centrale de Paris, 1996.

[HUE 63] HUET C., Étude par une méthode d'impédance du comportement viscoélastique des matériaux hydrocarbonés, PhD Thesis, Faculté des Sciences de l'Université de Paris, 1963.

[PER 03] PERRET J., Déformations des couches bitumineuses au passage d'une charge de trafic, PhD Thesis, École Polytechnique Fédérale de Lausanne, 2003.

PART 5

Railways and Inland Navigation

34

Application of Different Methods for Rehabilitation of Existing Transition Zones on Old Railway Lines

One of the basic objectives of the present railway authorities refers to the reduction of maintenance costs allocated for maintaining the railway infrastructure. Due to the frequent need for conducting additional maintenance and renewal (or even overhaul), which are disproportional compared to other sections of the railway track, several European railway authorities define the transition zones between the "normal" open tracks and "rigid" track sections or substructures such as bridges/tunnels/culverts as problematic parts on the track. This chapter presents a case study of the "Buna" bridge in Croatia, which focuses on the design and implementation of new innovative solutions for transition zones.

34.1. Introduction

One of the basic objectives of the present railway authorities refers to the reduction of maintenance costs allocated for maintaining the railway infrastructure. Due to the frequent need for conducting additional maintenance and renewal (or even overhaul), which are disproportional compared to other sections of the railway track, several European and American railway authorities define the transition zones between the "normal" open tracks and "rigid" track sections or substructures such as bridges/tunnels/culverts as problematic parts on the track [COE 10, AAR 93]. Frequent repairing of the places like these has resulted in reduction of the track capacity and traffic continuity, which resulted in additional costs to the railway authority.

Chapter written by Marko VAJDIĆ, Duo LIU, Stanislav LENART and Irina STIPANOVIĆ OSLAKOVIĆ.

Transition zones between bridges, tunnels, artificial and earth structures, including transitions between ballast and non-ballast permanent way (slab track), are a part of the railway track structure where the abrupt change in the rigidity of the track structure and track settlement occurs between individual transverse profiles, as a result of the change in the structural elements and the foundation. Regarding the emerging needs, higher axle load, increase in the traffic load and higher train speed, the transition zone issues are more urgent.

In order to limit additional and frequent costs of rehabilitation or renewal of the transition sections, the degradation mechanisms are analyzed within the SMART RAIL, FP7 research project, with the aim to find a high-quality, economically and environmentally acceptable solution for revitalizing deteriorated and aged existing rail infrastructure [GAV 12].

This chapter presents a case study of the "Buna" bridge in Croatia, which focuses on the design and implementation of new innovative solutions for transition zones between the open track on the embankment and the bridge. Starting with reviewing and evaluating the existing conditions by the application of geotechnical and geophysical investigation works, different approaches for both ends were applied for a systematic comparison of the overall performance. Innovative materials such as cementitious composites, geotextiles and geogrids were included in the design. Geotechnical-specialized FEM modeling work with PLAXIS software was then carried out for a preliminary analysis and evaluation of both approaches. Transitions were designed in a way that both settlements and stiffness variations are smoothened in a reasonable way so that both the safety and comfort of passing trains are not negatively affected at these locations.

34.2. Transition zones

Transition zones are defined as parts of the railway track where a change in basic characteristics that define a railway structure in its entirety takes place [ESV 01, JEN 06, ROS 08]. Under the basic characteristics, the following parameters are considered: substructure and superstructure stiffness, deformation of each substructure layer and each superstructure part, overall value of track deformations, geometric restraints. The transition zones in general represent the appearance of discontinuity in the track structure [UIC 08].

The type of the transition zone depends on where it is located:

– transition between different types of superstructure;

– transition from one type of substructure to another;

– direct transition between two different types of rigid substructures.

Within the SMART RAIL project, focus is on solving the problem in the transition zones between two different types of substructure, between the open track on embankment and the bridge.

34.2.1. *Negative mechanisms that occur in the transition zones*

Poor condition of the transition zones is a consequence of numerous complex and inter-related mechanisms. In order to find the best possible solution for solving the problems that occur in the transition zones, all the negative mechanisms that influence the behavior of the track structure should be taken into account and analyzed. Negative mechanisms that occur in the transition zones are as follows:

- discontinuity in the stiffness of the track structure;
- differential settlements of the rail track structure;
- influence of rail services speed;
- influence of the direction of the train.

The above-mentioned degradation mechanisms, which in fact each act on their own, may also be conditioned by each other. Cyclic repetition of these processes accelerates degradation of track geometry, with reduced quality and safety of driving as an immediate consequence [VAJ 12].

34.2.2. *Role of transition zones*

A certain structural solution is in fact hidden behind the term "transition zones". The main role of transition zones is to prevent sudden changes in stiffness of the load-bearing structural elements of the track. The aim is to minimize or prevent the occurrence of additional negative dynamic loads over a part of a transition zone, which additionally accelerate the track geometry degradation, with reduced quality and safety of driving as an immediate consequence. This can be achieved by linearly changing certain properties of the surrounding structures at a reasonable distance by dividing one differential change into smaller steps, i.e. dynamically irrelevant intervals [ESV 01, JEN 06].

Ideally, these inconsistencies occurring in parts of transition zones do not influence the performance of a passing train in terms of safety but rather they more often reflect upon the quality and comfort of rail services and other dynamic occurrences [ROS 08].

34.3. Case study: transition zones at the "Buna" bridge

The "Buna" bridge, situated at the railways track M104 Novska, Sisak, Zagreb at 398 + 422 km, is selected for the case study because of the obvious problems in the transition zones, immediately before and after the bridge. In the transition zones, irregularities in the geometry have been noticed, such as track unevenness and under ballast gaps, and vertical displacements of the whole track structure. These were caused by differential settlements and by dynamic impacts of the train due to the changes in track stiffness.

a) b)

Figure 34.1. *a) "Buna" bridge; b) deformed geometry of superstructure*

34.3.1. *Technical characteristics of the track section*

Longitudinal gradient on the planned intervention section is 0.02‰. According to the characteristics of this track, it belongs to the lowland tracks. The original project documentation for this section dates back to the 1893, when this track was designed and built. The line was for the first time rehabilitated between 1971 and 1972, implementing new materials in the superstructure according to the rehabilitation project dating from 1953. Then, the speed limit was 60 km/h, and allowed axle weight was 22.5 t/axle and 8.0 t/m. Next, rehabilitation that included the whole railway line from Novska to Zagreb in 2008 enabled maximum axle weight 22.5 t/axle and 8.0 t/m, with the speed limit of 140 km/h, but unfortunately the reconstruction skipped the "Buna" bridge and therefore at this section the speed limit is today restricted to 50 km/h. The original rehabilitation project for the "Buna" bridge replacement was designed in 2010, which was then extended with the new solution for transition zones in 2013.

According to the business plan of Croatian Railways (HŽ Infrastructure), reconstruction of the "Buna" bridge was planned for 2013, and that is also the reason why it was chosen for the case study.

Currently, rails are fastened on the wooden sleepers at a distance of 60 cm, 50 m before and after the "Buna" bridge with rigid fastening system (type K). The rest of the track has prestressed concrete sleepers. The wooden sleepers are much less stiff compared to the rigid concrete sleepers, and thus they easily absorbed additional dynamic effects that degraded the track structure. Their use in zones before and after the bridge and on the bridge presented one kind of solution that was used in the past for transition zones to prevent negative consequences in the track caused by the differential stiffness.

The new track design predicts the rails type 60E1 with adequate hardness (hardness category of at least R260) and inclination 40:1, elastic fastening system SKL-14 (direct fastening without a base plate), prestressed concrete monoblock sleepers, ballast of adequate quality and protective layers of the adequate thickness (minimum 30 cm) and material texture. Sleepers are installed at distance between centers of sleepers of 60 cm (1,670 pieces per km). Continuous welded rail will be used on the entire section.

34.3.2. *Traffic data*

At the selected section, there is mixed type of traffic; on average, there are 28 passengers and 13 freight trains daily, which gives about 1,270,827 passengers and 1,554,401 tonnes of goods annually. The types of locomotives that are operating on this railway line are as follows: 1141, 2132, 2041 and EMU 6111 [CRO 12].

34.3.3. *Performed geotechnical and geophysical investigation works*

An extensive geotechnical and geophysical investigation was been carried out on site in March 2012. Field tests have been supplemented by numerous laboratory tests to provide accurate information about ground conditions, as shown in Figure 34.2. All these investigations were performed in order to detect the causes of degradation in the transition zones and to provide necessary information for an appropriate reconstruction solution for that area.

The soil profile on the location of "Buna" bridge site was determined as follows:

– the railway embankment, which is about 1.0 m thick, consists of crushed stone and gravel material;

– the subsoil to a depth from 3.5 to 3.8 m consists of high plasticity clay (CH) of soft consistency, anorganic, blue-gray colored. At borehole B3 and B4, high-plasticity firm to stiff green-gray colored clay is to a depth 4.5 m;

– the following layer up to the depth 4.80- -5.70 m consists of silty sand (SM), medium to fine grained, loose, gray to dark brown colored. Dominantly high plasticity silt (MH), dark brown colored with soft consistency was registered at borehole B3;

– the bottom layer to the boring depth of 12 m consists of silty, poorly graded gravel (GP/GS) with gray colored rounded firm grains. Pockets of medium to fine grained silty sand (SM), green-gray colored, were found at boreholes B1 (from 8.4 to 9.3 m) and B2 (from 8.3 to 9.7 m).

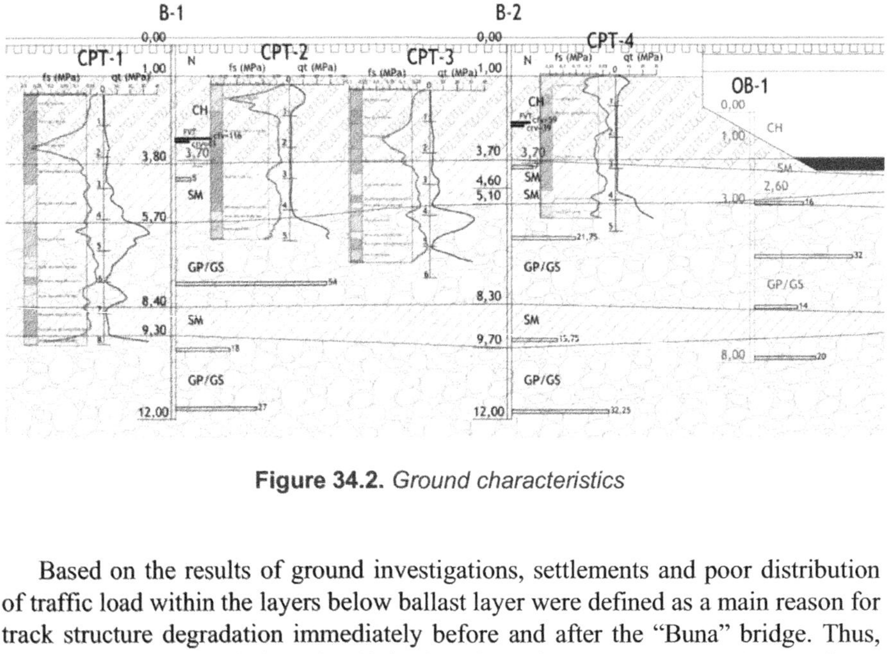

Figure 34.2. *Ground characteristics*

Based on the results of ground investigations, settlements and poor distribution of traffic load within the layers below ballast layer were defined as a main reason for track structure degradation immediately before and after the “Buna” bridge. Thus, the transition zone solution should be based on the strengthening of the railway embankment and the wedge behind the abutment.

34.4. Modeling

The idea was to design two different transition zones, from each side of the “Buna” bridge in order to have the ability to compare more than one solution. The first one (from Zagreb side) is based on present experiences described in the guideline “UIC CODE 719R: Earthworks and track bed on railway lines”

(Figure 34.3(a)). This technical solution is based on the application of various types of embankment materials (cement stabilized layers, infill material with controlled composition and granulation, material with better mechanical characteristics). The second one (from Sisak side) represents innovative approach using prestressed geogrid reinforced soil (GRS) (Figure 34.3(b)).

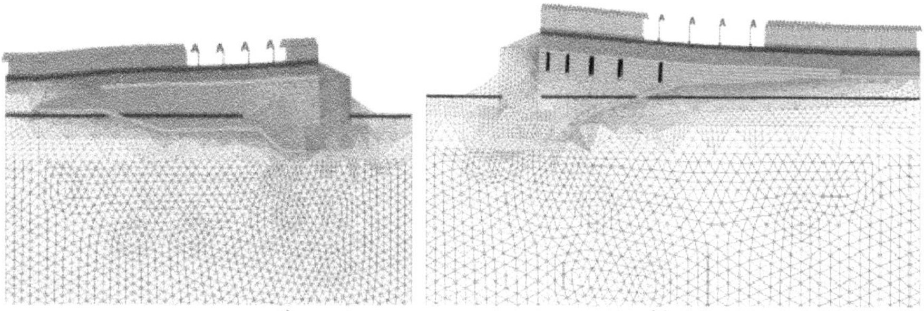

a) b)

Figure 34.3. *a) Transition zone on Zagreb site (No1); b) transition zone on Sisak site (No2)*

Calculations of load impact of railway embankment, replacement layers in the transition zones and the foundation soil (stress distribution in the soil, and distribution of deformations and displacements) were performed using Plaxis 2D (version 2010.01) and Plaxis 3D Foundation (version 2.2) based on the finite element method. The soil is modeled using a nonlinear hardening model, hardening small strain stiffness model and the Mohr–Coulomb model.

The computational model has been developed based on the geometry of the rehabilitation project and the layers and characteristics of the soil. It is composed of two-dimensional and three-dimensional finite element mesh. The boundaries of the model, taking into account the distance from the highest strain changes, have been selected according to the usual rules of numerical modeling. In the nodes at vertical boundaries, horizontal displacements are disabled, whereas in the nodes at lower boundaries, vertical and horizontal displacements are disabled.

Calculations were carried out with effective strength parameters taking into account the non-drained condition. Computational soil models have been adopted on the basis of investigation data presented in the previous section. Assumed deformation characteristics of geosynthetic reinforced soil were based upon the results of laboratory tests [LEN 12] (Figure 34.4).

Figure 34.4. *Laboratory tests of GRS. For a color version of this figure, see www.iste.co.uk/jacob/materials.zip*

34.4.1. *Calculations with Plaxis*

After the calculations of initial soil strains, calculations of deformations were conducted in the following phases:

1) activating substitute materials in the transition zone;

2) activating the traffic load according to the UIC71 scheme at the beginning of the transition zone;

3) moving the load for 0.2 m in 2D model and for 1.0 m in 3D model, or activating loads at the displaced location all the way to the abutment;

4) defining the load at the end of the model at the abutment.

Load according to the UIC71 scheme is approximated. The overview of calculation results for the existing (past) condition at the transition zones and new designed conditions at both sides of the bridge are given in Table 34.1.

34.4.2. *Interpretation of the results*

On the basis of a performed analysis of the distribution of load and settlement calculations, it can be concluded that the offered technical solutions, if compared with the current situation, are enabled to achieve gradual linear change in stiffness of the bearing substructure in the transition zone, and a better distribution of traffic load through the track structure itself, thus reducing the value of total settlement of the same.

Software application Plaxis 2D model		Calculation stages (display of vertical displacement u_y) (m)		
		Initial stage of the moving load (the load is dominant at the end of TZs)	A central phase of the moving load (the load is dominant at the middle of TZ)	The last stage of the moving load (the load is dominant at the beginning of the TZ – along the abutment)
Vertical displacement at the beginning of the TZ (along the abutment)	Existing TZs	0.013	0.014	0.029
	TZ no. 1	0.004	0.005	0.006
	TZ no. 2	0.006	0.007	0.009
Vertical displacement at the end of the TZ	Existing TZs	0.031	0.029	0.024
	TZ no. 1	0.015	0.013	0.012
	TZ no. 2	0.021	0.018	0.017

Table 34.1. *The calculation results of the transition zones (TZs)*

34.4.3. *Final design: direction towards Zagreb*

The transition zone is formed behind the abutment wall under the ballast layer. Its final length is 17 m (Figure 34.5). The embankment stabilized by hydraulic binder is proposed behind the abutment. The embankment is built from the bottom of the abutment foundation to the top of the abutment. The slope of the stabilized part of the embankment is 1:1. The subgrade material of controlled composition and certain degree of compaction is built with a slope 1:2 into the embankment to the end of the transition zone.

A layer of geotextile is going to be laid at the bottom of the construction along the transition zone. Another layer of geotextile and a geogrid is laid on the surface of the formation layer. A 30-cm thick cement stabilized base layer is proposed at the top of the formation layer near the bridge at the length of 4 m.

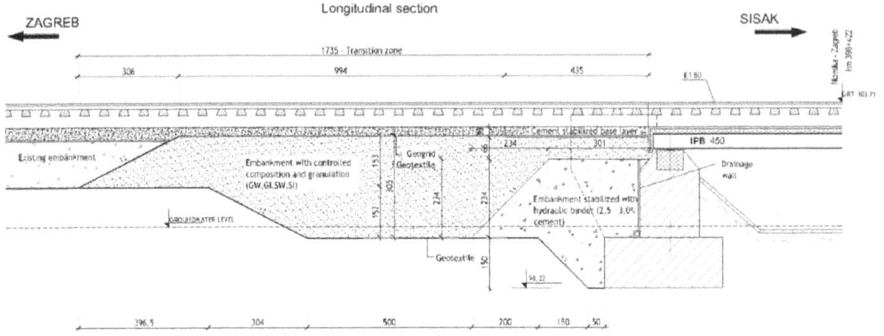

Figure 34.5. *Final design of the transition zone on Zagreb site*

34.4.4. *Final design: direction towards Sisak*

A prestressed GRS is going to be used in transition zone between the "Buna" bridge and open ballasted track in the direction of Sisak. The length of transition zone is also 17 m. The subgrade material of controlled composition and a certain degree of compaction is built into the embankment from bridge abutment to the end of the transition zone. The embankment is constructed in a slope 1:1 from the bottom to the top of the abutment foundation and in a slope 1:2 to the depth 1 m below the top of the embankment. A layer of geotextile is to be laid at the bottom of embankment. The upper 1 m of embankment is constructed by GRS. Horizontal layers of geogrid have a vertical spacing of 20 cm. GRS starts behind the abutments and continues from 8 m (at the bottom) to 17 m (at the top) in the direction going away from the bridge (Figure 34.6). Subgrade material of controlled composition used for the embankment construction in the transition zone is used as a backfill material in the upper 40 cm of GRS, while the bottom 60 cm of GRS is formed from the backfill material that is used for construction of embankments on the open track. In order to obtain higher stiffness of the formation layer, GRS is designed to be prestressed in vertical direction at the length of 8 m, direction going away from the bridge.

34.5. Construction process

The construction process was started in June 2013 and it was expected to end in November 2013. During the construction works, it was not possible to close the track permanently. The new bridge design involved the replacement of the old steel bridge with the new prefabricated concrete bridge. During the execution of substructure and abutments rehabilitation works, provisional (temporary) bridge was made. Due to all those circumstances, certain modifications in the design of the transition zones had to be performed.

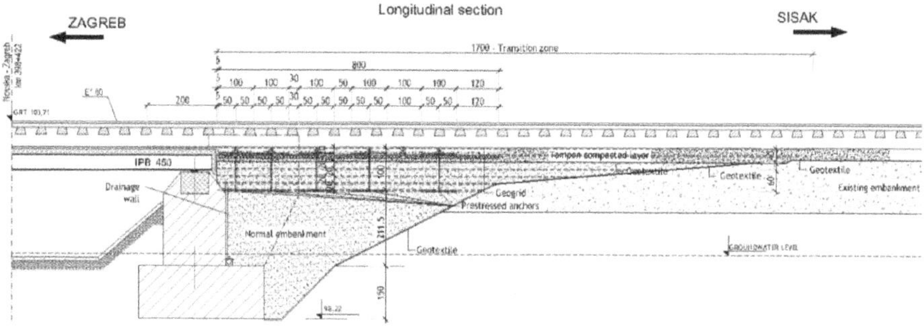

Figure 34.6. *Final design of the transition zone on Sisak site*

Before completing the work, the transition zone on the Sisak side of the bridge was successfully constructed and implemented. Prestressing of GRS was performed by the application of anchor plates of dimensions 0.8 m × 0.8 m × 0.02 m on both sides of prestressed vertical steel rod (IBO anchors) to transfer reaction load on the GRS, as shown in Figure 34.7. IBO anchors had diameter $\Phi = 32$ mm.

Figure 34.7. *Execution of GRS layer 1*

The maximum prestressing force was 250 kN. Anchors were prestressed in accordance with the scheme shown in Figure 34.8.

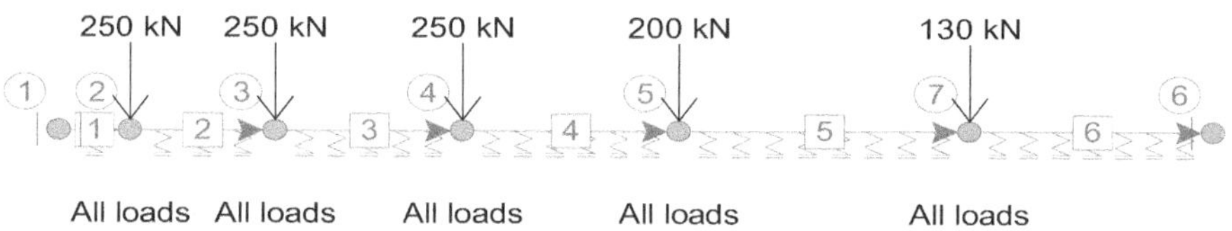

Figure 34.8. *Load scheme of anchoring longitudinal beam at the top of GRS*

In order to optimize the distribution of prestrssing forces in the longitudinal direction of transition zone, GRS layers are connected with two steel beams placed in longitudinal direction as shown in Figure 34.9.

Figure 34.9. *Execution of GRS layer 2*

34.6. Monitoring

During the construction works, the level of compaction was constantly measured as well as the quality of embedded materials. After the construction of the transition zones, it is necessary to carry out monitoring and measurements of deformations and stresses in order to calibrate the computer model and confirm the proposed solutions. Therefore, extensometers have been embedded (welded) on the geogrids, in a total of 72 strain gages on 13 different locations, and on geotechnical anchors, in a total of 10 strain gages on five different locations. In order to have continuous monitoring, horizontal and vertical tubes have been implanted to enable settlement measurements using an inclinometer (Figure 34.10).

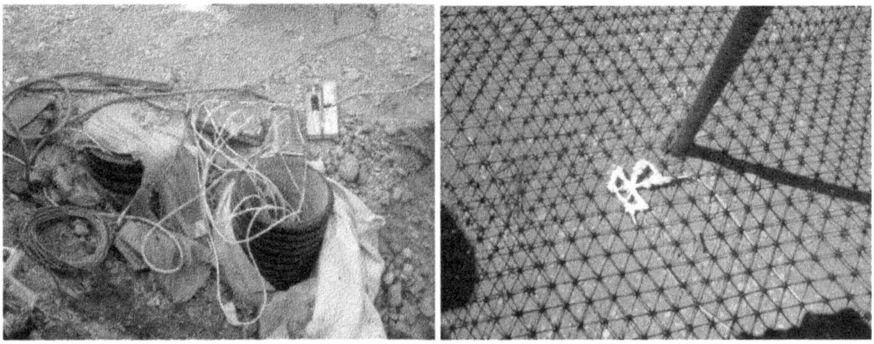

Figure 34.10. *Strain gages*

34.7. Conclusions

Transition zones represent discontinuity in track structures. Many studies have indicated that sudden changes in track structure stiffness in transitions over culverts or transitions from an open part of the route onto a bridge/viaduct/tunnel are the main reason for the degradation of track structure in those sections. The higher the axle load is, as the traffic load increases and the train speed is higher, the more urgent the transition zone issue is.

There are currently no specific recommendations for dimensioning and design of the transition zones in Croatia where many projects for the rehabilitation and improvement of existing railways are planned or already ongoing.

With this project, for the first time in Croatia, transition zones have been implemented in practice, as previously dimensioned and engineered "structures", as a part of the whole railway track design.

By improving durability properties of the track structure in the transition zone, we contribute to the reduction of future costs intended for maintenance, which brings a direct benefit not only to infrastructure owners but also to operators and end users.

34.8. Acknowledgments

This research is performed within FP7 EU research project "Sustainable Maintenance and Analysis of Rail Transport Infrastructure (SMART Rail)", supported by EC funding, under the program SST.2011.5.2-6. TPT with the aim of cost-effective improvement of rail transport infrastructure.

34.9. Bibliography

[BAR 05] BARR M., Identification of the best practices for design, construction, and repair of bridge approaches, Center for Transportation Research and Education, Iowa State University, Iowa, 2005.

[COE 10] COELHO B. HOLSGHER P., PRIEST J. *et al.*, "An assessment of transition zone performance", *Journal of Rail and Rapid Transit*, vol. 224, 2010.

[CRO 12] CROATIAN RAILWAYS, Annual report 2012, Zagreb, 2012.

[ESV 01] ESVELD C., *Modern Railway Track*, second ed., TU Delf, 2001.

[GAV 12] GAVIN K., OSLAKOVIĆ S.I., VAJDIĆ M. *et al.*, "Smart maintenance and analysis of railway transport infrastructure (SMART RAIL)", *Second International Conference on Road and Rail Infrastructure (CETRA 2012)*, Dubrovnik, Croatia, pp. 421–428, May 2012.

[JEN 06] JENKS C.W., Research Results Digest 79, Washington, DC, 2006.

[LEN 06] LENKE L.R., "Settlement issues – bridge approach slabs (final report phase I)", University of New Mexico, Department of Civil Engineering, New Mexico, 2006.

[LIC 05] LICHTBERGER B., *Track Compendium*, Eurailpress, Hamburg, 2005.

[PER 10] PERIĆ Z., POPOVAC N., Integralni mostovi, Zagreb, 2010.

[PLO 06] PLOTKIN, "Track transitions and the effects of track stiffness", TTCI/AAR, 2006.

[ROS 08] ROSSMANN N., *Transition Areas: Ballasrless – Ballasred Track*, Neumarkt, 2008.

[TRA 93] TRANSPORTATION RESEARCH BOARD, Settlement of Bridge Approaches (The "Bump" at the End of the Bridge), Transportation Research Board, Washington, DC, 1993.

[UIC 08a] UIC CODE 719R, *Earthworks and Track Bed on Railway Lines*, UIC, Paris, 2008.

[UIC 08b] UIC, *Vertical Elasticity of Ballastless Track*, UIC, Paris, 2008.

[VAJ 12] VAJDIĆ M., OSLAKOVIĆ S.I., KRALJ S., "Transition zones on the railway track – overview", *Second International Conference on Road and Rail Infrastructure*, Dubrovnik, Croatia, pp. 633–640, May 2012.

35

CAPACITY4RAIL: Toward a Resilient, Innovative and High-capacity European Railway System for 2030/2050

CAPACITY4RAIL is a European Union funded research project that was started at the end of 2013. Under the coordination of UIC, it brings together railway operators, infrastructure managers, manufacturers, world leading research organizations and specialist technology providers. It will deliver research that is innovative, prepares rail for the future and takes into account results from previous research projects and programs. The project is based on previous useable results and will deliver both technical demonstrations and system-wide guidelines and recommendations that will be the basis for future research and investment, increasing the capacities of rail networks in the future. The project aims at developing new concepts for low maintenance and resilient infrastructure as well as optimizing operation and intermodal integration within the global transport system.

35.1. Introduction

In 2011, the White Paper on European Transport reasserted how fundamental transport was for society, for the mobility of European citizens and for the growth and vitality of the European economy. CAPACITY4RAIL is a European Union (EU) funded research project that was started at the end of 2013. Under the coordination of UIC, it brings together a wide range of stakeholders in an ambitious partnership that sees railway operators, infrastructure managers, manufacturers, world leading research organizations and specialist technology providers. It will

Chapter written by Laurent SCHMITT, Fabien LÉTOURNEAUX, Isabelle DE KEYZER and Paul CROMPTON.

deliver research that is innovative, prepares rail for the future and takes into account results from previous research projects and programs. The project is based on previous useable results and will deliver both technical demonstrations and system-wide guidelines and recommendations that will be the basis for future research and investment, increasing the capacities of rail networks in the future. The time used for infrastructure monitoring, maintenance and renewal means "down time". New concepts for low maintenance infrastructure using standardized and "plug-and-play" concepts will be proposed. Non-intrusive innovative monitoring techniques or self-monitoring infrastructure will be investigated, allowing low or no impact on train operations. The fragility of some key components of the infrastructure system (especially in extreme weather conditions) such as switches may impact the efficiency of the whole system. The resilience of switches to any kind of known failure will be reinforced, as well as the ability of the operation system to recover from incidents. Capacity enhancements will also be achieved by higher speed freight vehicles, allowing an optimized interleaving of freight trains into mixed traffic, and improved planning models for operation. Intermodal integration within the global transport system will be improved through enhanced transshipment of passengers and freight. CAPACITY4RAIL will also look toward 2030/2050, by proposing guidelines for future deployments in the mid-term, recommendations for technologies to be developed and deployed in the long term and investigating the key opportunities for funding these within national and EU funding schemes.

In order to best address this approach, the project has been divided into six subprojects (SPs). These are as follows:

- SP1 infrastructure;
- SP2 new concepts for efficient freight systems;
- SP3 operations for enhanced capacity;
- SP4 advanced monitoring;
- SP5 system assessment and migration to 2030/2050;
- SP6 management, dissemination, training and exploitation.

35.2. Project objectives

The overall objective of CAPACITY4RAIL is to increase capacity, availability and performance of the railway system through major step changes in:

- infrastructure design;
- construction and maintenance (including advanced monitoring);

– operations management;

– incident recovering through real-time data management;

– freight operations, with a particular focus on transhipment and improved specifications for rolling stock.

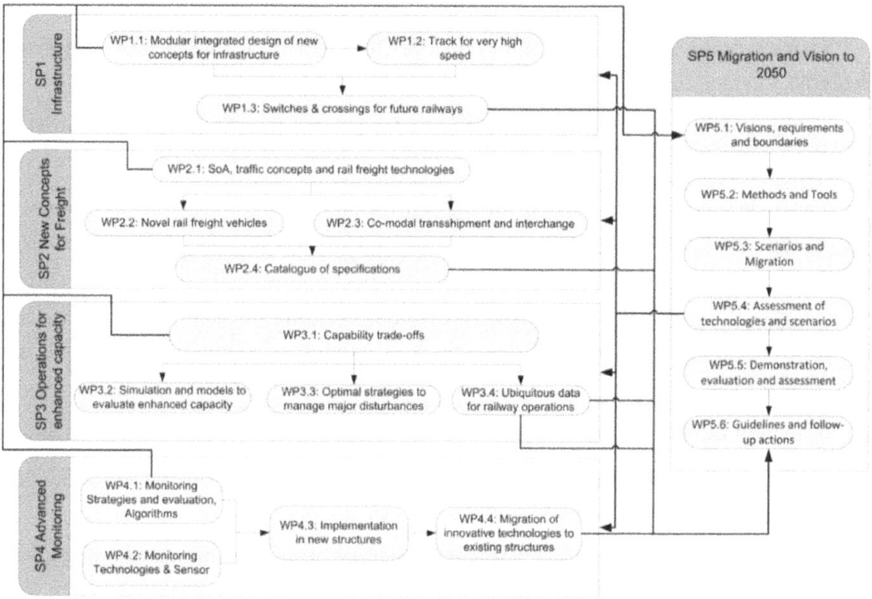

Figure 35.1. *CAPACITY4Rail structure and interactions*

Detailed objectives for each of the SPs are listed in section 1.1.5 "Project Approach", but CAPACITY4RAIL will:

– provide an overall increase in railway capacity by developing a holistic view on the railway as a system of interacting technical components driven by customer demand;

– develop the vision and requirements of the railway system in 2030/2050 including environmental and socioeconomic aspects in terms of infrastructure and operation (including provision for an increase of freight traffic);

– identify the technologies and development/implementation steps in short, medium and longterm, necessary to move toward the railway system for 2030/2050;

– develop new concepts for railway track of the future, in view of potential application for mixed traffic, but also very high speed (over 350 km/h), that encompasses cost savings, rapid construction, resilience and enhanced maintainability;

– provide general track design guidelines for very high speed with identification of limiting factors, especially in terms of admissible track irregularities and transition zones;

– demonstrate, in real rail freight corridors, the feasibility of the most promising designs;

– specify the requirements of an efficient freight rail freight system 2030/2050 to fulfill the EU targets;

– provide a conceptual design for the rail freight vehicles (wagons and trains) of the future design transhipment technologies and interchanges of the future (rail yards, intermodal terminals, shunting facilities, rail-sea ports, etc.);

– design modern fully integrated rail freight systems for seamless logistics and network-based performance and develop their catalogue, standardization pack and technical spec;

– develop modeling and simulation tools for high volume traffic management;

– derive joint requirements, European standards and testing for incident management plans;

– develop new concepts for railway structural and operational monitoring to enhance the availability of the track;

– elaborate a system for a fastcheck of track and structures after natural hazards.

Each of the proposed SPs will have an effect on its corresponding domain in the mid to long term. The following is a summary of the key performance indicators that each SP expects to impact:

– *Infrastructure*:

- lifecycle costs (LCCs) for slab track;

- availability for capacity: less time spent for intervention including maintenance and repair following unpredicted events.

– *Freight*:

- carrying capacity per train;

- number of trains per track length;

- investment cost per train;

- maintenance cost per train;

- cost per produced ton × km;

- reliability increase.

– *Operations*:

- capacity from consistently predictable train movements;

- reliability for passenger and freight;

- more efficient response to perturbation;

- more efficient use of infrastructure, energy and rolling stock;

- reduction in operational costs through manual intervention.

– *Monitoring*:

- reliability of infrastructure because of early pre-failure detection;

- availability of infrastructure because of less failure and non-intrusive monitoring;

- higher maintainability of infrastructure because of earlier and more accurate diagnoses.

35.3. Project approach

35.3.1. *Infrastructure*

The efficiency of current railway operation in Europe is often lowered by incidents that cause delays, or major engineering works on the infrastructure that entails either delaying or rescheduling trains. Significant progress in terms of track availability for running trains and increased resilience will come only from new track systems that are designed to provide greater reliability through focusing on designs targeted to the actual use and much lower maintenance. An improvement of the reliability of switches and crossings is also required.

The SP infrastructure will deliver the following:

– new concepts of track mainly based on modular slab track incorporating as far as possible embedded elements for power system, signaling and telecommunication, in view of mixed traffic but also for very high speed;

– new concepts for switches and crossings designed based on failure modes analysis, revisiting curving physics and incorporating sensors for condition monitoring;

– new optimized designs for switches that are more resilient to extreme weather condition;

– prototypes of the most promising concepts both for regular track and switches demonstrated on mixed traffic lines;

– new reliability, availability, maintainability, safety (RAMS) and the cost-driven track design method;

– improved modeling knowledge of track and subgrade behavior to severe thermal and humidity condition, along with the design guidelines with respect to flood and heavy rain;

– general track design guidelines at very high speed and identification of limiting factors, especially in terms of admissible track irregularities and transition zones.

– improved knowledge on the dynamical behavior of very high speed trains on bridges and appropriate design principles.

35.3.2. *New concepts for efficient freight systems*

The main goal is here to develop a modern, high-performing, automated, intelligent, fully integrated system for efficient, reliable and profitable freight operations capable of providing computerized real-time information to customers and operators at any time. A virtual simulation, working as guidance for real demonstrators, adapted to the real situation in different European regions, will be produced and disseminated to the various actors within the European transport and political sector. The newly designed systems will be market driven ensuring seamless low-cost operations for service providers and customers alike. The main idea is based on the concepts, ideas and results of previous projects and channels their merits into a logistics design that maximizes the system performance of a future supply chain seamlessly integrating rail with maritime and road transport. This means introduction of economically feasible technical development of rolling stock, capacity management, terminal operations, infrastructure maintenance and upgrading. New traffic management systems for better utilization of existing network capacity can only be fully exploited together with better performing rolling stock.

The results will substantially increase the overall capacity of the rail freight system, with adapted time- and cost-efficient terminals, multiple innovations on the wagons, and simultaneously introduce new modernized wagon components and novel sensors and monitoring, also using the enhanced traffic operation tools – thereby allowing freight trains to easily blend with passenger train traffic. The target is an overall capacity improvement by 5–10% at the end of the project, 20% in 2020 and an ability to guide how to proceed to a 50% increase by 2030. Prestudies on new concepts and designs of rail freight systems and networks have now been carried out; they indicate that these improvements can be reached.

The results will be specified for the different types of stakeholders and will specify system-related measures for all the following areas:

1) train/wagons: length, speed, performance, central/automatic couplers, EP/electronic braking, electrification, automation and weight;

2) operations: wagon shunting, intelligence for vehicles in terminals and terminal operation;

3) facilities: marshalling yards and terminals.

35.3.3. *Operations for enhanced capacity*

Railway operations strategies that will increasingly use automation for optimized performance and enhanced capacity will be developed. This SP will develop road maps for technology that will transform decision support systems into automated systems that enable the rail industry to meet the challenges of the future, such as high-speed freight and greater levels of transhipment between rail and other modes.

This will enable the future controllers of the railway to focus on fulfilling the challenges of running "on demand" train variations in resource allocation and collaborative working within and between countries and improved efficiency of trans-shipment at nodes while the systems take care of routine operations and recovery from small or maybe even medium perturbations, which in today's world need the intervention of the controllers.

The SP will issue the following:

– approaches, the majority of which will be fully automated, that help planners to understand and prioritize system capabilities and decide on optimal strategies to increase overall system capability, respond dynamically to planned and unplanned changes, and support real-time punctuality management. These strategies will take into account the requirements from SP2 for high-speed freight operations;

– new and innovative data processing methodologies and algorithms to interpret and summarize raw measurements to make the data more practical for use in data analysis tools;

– more relevant and effective data analysis tools to enable the processed data to be used in decision making, and new data visualization and presentation tools to assist strategic and operational decision making;

– research that focuses on data systems that are able to provide ubiquitous data on train position and condition that enable automatic decision support systems and operations and planning staff to make better decisions;

– proof of concept models that support the development of roadmaps for future modeling and simulation and operational strategies that provide a vision for future improvements in automated railway operations;

For each of the outputs, improvements will be quantified through simulation or real-world demonstration of the project outcomes.

35.3.4. *Advanced monitoring*

The objectives of advanced monitoring are to develop new concepts for railway structural and operational monitoring to enhance the availability of the track combined with automated maintenance forecasts, a prediction of the structural lifetime, a fastcheck of track and structures after natural hazards and a support for train operation by train monitoring.

The expected results of this SP will define railway-specific monitoring strategies to improve RAMS and LCC of railway transport, provide an overview of up-to-date monitoring technologies from other industries, develop a concept for further use of information and communication technologies in maintenance and for operation support, new concepts for using advanced monitoring in embankments, bridges, different tracks types, switches, etc. designed for life-time prediction, condition-based maintenance and fast health reporting after natural events (e.g. extreme weather), as well as prototypes of the most promising concepts both for regular track and switches demonstrated on mixed traffic lines. They will include recommendations for approved, sustainable sensors and energy harvesting technologies in railway applications and will provide improved modeling knowledge of track and subgrade behavior for more efficient application of monitoring systems.

35.3.5. *System assessment and migration to 2030/2050*

A common task will run through all of the technical ones to examine the state of the art, evaluate existing data and customer requirements and define the requirements and boundaries for the work to be carried out. It will harmonize these approaches and deliver the overall system requirements.

Global scenarios will be then developed for future rail systems and at the end of the project will bring together the final results, analysis and recommendations ensuring the future work will fit into this same system approach and ensuring the migration from existing system to the future system. It will describe the common vision for an affordable, adaptable, automated, resilient and high-capacity railway, and develop a "roadmap" that paves the way for the target system.

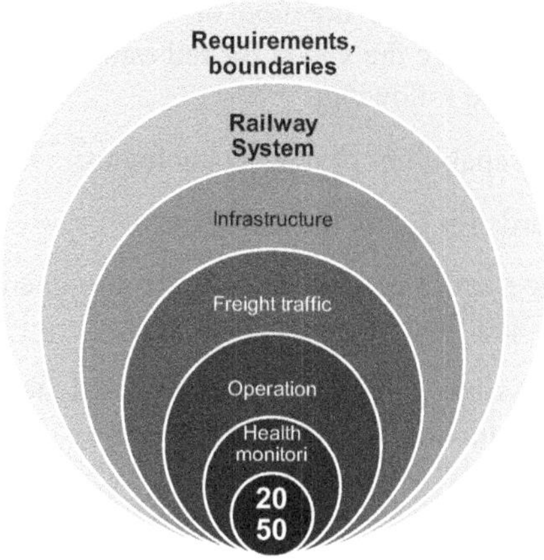

Figure 35.2. *Capacity4Rail migration model*

The results will be based on the outcomes from workshops, reviews of White Papers, policies and strategies of IMs. Also the experience from other transportation modes will be merged with the requirements for improvements developed in the other work packages.

Depending on the assessment of the "technology readiness level" (TRL) of the technologies to be developed, the direction and paths for migration from the existing system (i.e. what is the current situation today) to the visionary system (i.e. where we want to be) will be defined. The project will aim to identify and develop the specifications and technologies and also the interim steps that will resultin the increase in TRLs for key technologies. It will thus identify the milestones that are needed to be achieved in order to establish concrete step changes to achieve a European railway network of enhanced capacity.

Furthermore, supporting real and virtual demonstration of technologies, the SP will:

– evaluate the impact of the step changes in the new concepts on RAMS, LCCs, operation and environment;

– provide adapted methods and tools for the assessment of innovations, technologies and concepts with respect to the vision and roadmap as well as technical, economic and environmental requirements and boundaries.

35.4. Conclusions

Started in October 2013 for a duration of 4 years, CAPACITY4RAIL has a unique opportunity to influence the vision for rail transport over the next 30+ years. This will be achieved by the following:

– demonstrating the capabilities of new technologies;

– developing standards for future technologies;

– providing guidance and recommendations for the overall rail system and the implementation of new and future technologies;

– identifying all possible sources for funding and development that can be leveraged to ensure the future of rail transport in Europe.

Of course, to achieve these ambitious objectives, the consortium is required to have a diverse range of stakeholders who "buy in" to the CAPACITY4RAIL vision and adopt the system approach that is advocated in the project. CAPACITY4RAIL has brought together this diverse range of stakeholders, in an ambitious partnership that includes46 manufacturers, railway operators, world leading research organizations and specialist technology providers. The construction of the consortium has been driven by the technological needs of the overall project and the expertise required in each of the SPs.CAPACITY4RAIL is being coordinated by UIC which are in a unique position in the European Rail network and are ideally placed to lead such a complex project.

35.5. Acknowledgments

The research project is funded by the European Union Seventh Framework Programme under grant agreement no. 605650.

36

Secondary Stiffness of Fastening Clips: Influence on the Behavior of the Railway Track Panel

Fastenings are all the materials whose function it is to structurally connect the rail to the sleeper. The fastenings used in concrete sleepers must be elastic in both tension and compression. The basic component of all types of fastenings consists of an elastic steel rod or plate, named spring clip or spring plate. The clip functions as a system with a "compatible" elastic pad. The clip, as a spring, develops the elastic clamping force of the fastening on the rail foot. The load–deflection curve of the clip characterizes the static stiffness of the clip and consequently of the fastening. Some clips present only one type of static stiffness, but the most advanced present a secondary static stiffness also. In resilient fastenings, apart from the clip – acting on the upper part of the rail foot – there is a second spring called the resilient pad, which is made of elastic material compatible with the fastening steel clip, acting under the rail foot. The present study examines the influence of the secondary stiffness of the clip on the behavior of the rail in the track panel in terms of the rail tilt/rotation, the clamping force and rigidity.

36.1. Introduction

Fastenings are all the materials whose function is to structurally connect the rail to the sleeper. The fastenings used in concrete sleepers must be resilient, presenting elasticity in both tension and compression, upward and downward. The basic component of all types of fastenings consists of an elastic steel rod or plate, called a fastening clip or fastening plate, which are springs. One pair of clips per rail, acting on the upper part of the rail foot, together with one "compatible" resilient pad,

---

Chapter written by Konstantinos GIANNAKOS.

underneath the rail's seating surface, function as one ensemble. The pad provides elasticity (springs) under the rail's seating surface (Figure 36.1); it is produced from elastic material and it must be compatible with the clip. The compatibility is assured through the load–deflection curves of both the clip and the elastic pad. The clip's load–deflection curve must be relatively flat so that the clamping force remains adequate over the years. Some clips present only one type of static stiffness, but the most advanced also develop a secondary static stiffness. The clip, as a spring, develops an elastic clamping force on the upper surface of the rail foot, the toeload, due to its tightening after mounting (Figure 36.1). After the mounting of the clip, the pad develops a subsidence δ_0 and the clip's contact point a subsidence Δ_0. The load–deflection curve of the clip characterizes the static stiffness coefficient of the clip, determining:

– the clamping force on the rail foot in relation to the subsidence of the upper surface of the foot of the rail at the contact point of the clip on it, as in Figure 36.1, due to the tightening of the clip. The final position of the clip's contact point on the rail foot is replaced for Δ_0 and the pad is compressed for a deflection δ_0. As the axles pass over the support point (sleeper), the running loads increase the initial δ_0 and thus the initial Δ_0 decreases;

– the reduction in the clamping force in relation to the wear of the contact area between clip and rail, the fatigue of the pad or the loosening of the fastening system (due to the decrease in the initial Δ_0).

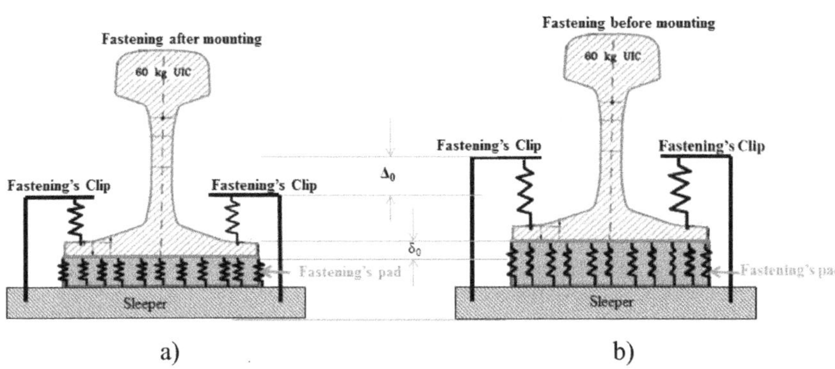

Figure 36.1. *Fastening in schematic depiction, before mounting (right side) and after mounting (left side)*

It is pointed out that for high-speed as well as for very-high-speed tracks, the fastening is the main element of advanced technology, and especially the clip and the elastic pad, which must be compatible. The present study – performed by the author as coordinator of the research team in Greece, for the dimensioning/

equipment of the high-speed track of the Hellenic Railway network (OSE), additionally to parallel investigations – examines the influence of the secondary stiffness of the clip on the behavior of the track panel in terms of the rail tilt/rotation, the clamping force and the rigidity of the track panel. The theoretical research results are compared to laboratory tests and *in situ* measurements. During the research on the resilient fastenings, in the Greek railway network (since the late 1980s and during the 1990s and later), three resilient fastenings on concrete sleepers have been examined: the Nabla fastening, the W14 fastening and the fast clip.

36.2. The railway track panel

The railway track is a multi-layered structure consisting of a vertical succession of various materials or layers of materials that define the final position of the rail running table as well as the properties of the track itself, as it "reacts" to the "action" that is created from the motion of the railway vehicle. Each material or layer that constitutes the line can be simulated by a combination of a spring with spring constant k_i and a damper with damping coefficient c_i. The "weak links" in this multi-layered construction are the ballast and the substructure. These develop residual deformations, deflections and lateral displacements, proportionally related to the deterioration of the geometry of the track, also referred to as quality of the track. The rails, sleepers and fastenings with their pads constitute the track panel. Since the permanent deformation of the track is a percentage of the total deflection resulting from the train circulation, it is imperative to reduce primarily the deflection of the layers, as much as possible, and also the lateral displacements. According to the theoretical analysis of Winkler [WIN 67], Weaver *et al.* [WEA 90] and Zimmermann [ZIM 41], the deflection of the track as an infinite beam on elastic foundation should be high enough to distribute the acting load to a longer section of the track and thus to reduce the reacting force at each point. This amount of deflection can be provided by a resilient fastening and its rail pad. The smaller the residual deformations and their increase over time, the better the quality of the track. The AASHTO testing for road construction equation for maintenance costs is also applicable for a railway track [GIA 04, EIS 04, EIS 97]:

$$(\text{decrease in track geometry quality}) = (\text{increase in stress on the ballast bed})^m \quad [36.1]$$

where $m = 3$ to 4.

The decrease in track geometry quality affects proportionally the maintenance costs and it is related to the stresses on the ballastbed and the degree of fouling of the ballastbed. The latter influences the preservation of the track geometry. Since stress is equal to the ratio of the actions on the sleeper (reaction per "point") to the seating surface of the sleeper, and the seating surface of each sleeper type is standard, a more precise estimation of the actions on the track should be performed in order to calculate the stresses (and consequently the maintenance cost) on the ballastbed. Professor Eisenmann [EIS 97, p.102] states that, on a track used by mixed types of traffic including intercity (IC), the effect of freight trains on track bed change is to be taken into account. According to the investigation carried out using the second and fourth power rules, by applying the second power rule (in equation [36.1]), with reference to the track bed change, 2.7 IC is equivalent to one freight train; by applying the fourth power rule, 3.5 IC corresponds to a freight train, concluding that this validates the inclusion of freight trains in the evaluation.

36.3. The system "railway track–railway vehicle" as an ensemble of springs and dashpots

Studies of "track analysis" for the definition of the actions on the track and its constituent components have led to the examination of the ensemble "vehicle-track" system (Figure 36.2), consisting of the vehicle with all its suspensions (primary and secondary), the car body and all the rolling instruments (axles or bogies) and the track, comprising of the "rail sleeper fastening-ballast-substructure". A subsystem of the latter system constituting the railway track is the "railsleeper" with its fastenings ballast, which, in the classical terminology, is the superstructure of the track (today, the blanket layer/subballast is included in the superstructure).

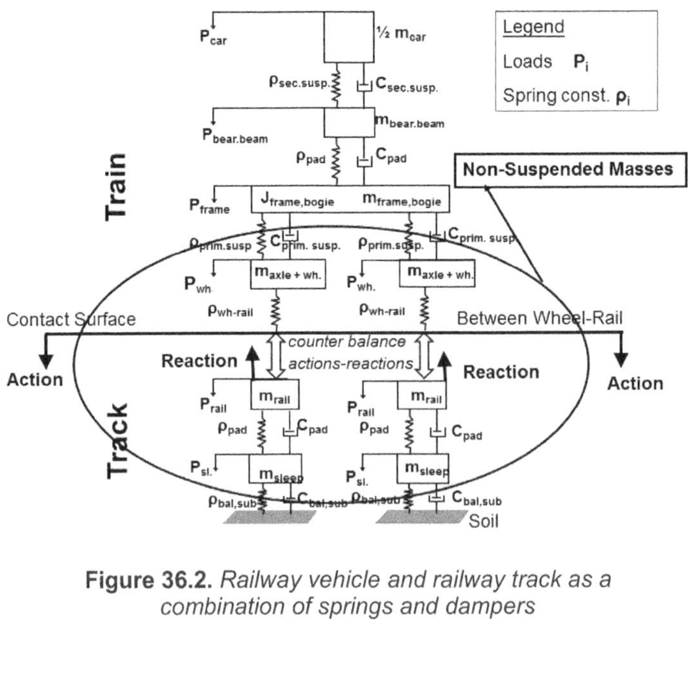

Figure 36.2. *Railway vehicle and railway track as a combination of springs and dampers*

The system "track" generally consists of five components (Figure 36.3) of spring constant ρ_i. If any of the five components is changed, the rest are affected and the whole system reacts differently. During the planning of a high-speed line (cross-section of a ballasted track and its simulation to an ensemble of springs and dashpots in Figure 36.3), all the details and the specifications for the five parameters should be followed – with no exceptions – otherwise significant problems to the overall functioning of the system may arise, with serious consequences either to the safety of traffic or to the cost of operation and maintenance of the track. Track maintenance is affected by the loading conditions of the vehicle motion combined with the constitutive layers of the track and their stiffness coefficients. This combination determines the acting reaction/action on each discrete point of support (sleeper) of the rail and the implied stresses on the ballast bed and the substructure. The stress on the ballast bed is defined as the ratio of the acting reaction divided by the seating surface of the sleeper.

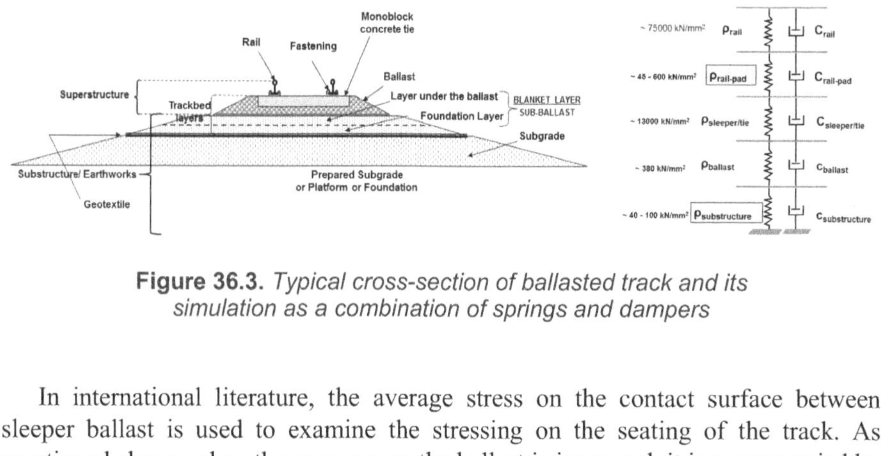

Figure 36.3. *Typical cross-section of ballasted track and its simulation as a combination of springs and dampers*

In international literature, the average stress on the contact surface between sleeper ballast is used to examine the stressing on the seating of the track. As mentioned above, when the pressure on the ballast is increased, it is accompanied by an increase in the maintenance cost due to the deterioration of the track geometry. During the study for the dimensioning as well as the selection of the individual materials comprising a railway track, the "weak links" are the ballast and the substructure. As already briefly described, these are the elements of the track that present residual deformations: subsidence and lateral displacement, directly connected to the deterioration of the so-called geometry of the track. For a given quality of ballast material, as far as the part of the deformations caused by the ballast and the earthworks of the track are concerned, this is accomplished by the correct combination and usage of heavy track machinery (ballast regulator, tamping machine, dynamic stabilizer). For the layers underneath the ballast, a very well-executed construction is required: crushed stone material in the upper layer, 100%

modified proctor or 105% proctor compaction. This specification implies a substructure almost undeflected [GIA 10a].

Taking into account the need of significant deflection in order to achieve a distribution of the acting load to the adjacent sleepers, this deflection should be offered by the fastening and its rail pad, demanding high resilience (small static stiffness coefficient) and consequently larger pad thickness.

36.4. The static stiffness coefficient of the railway track in the vertical direction

The stiffness (elasticity) coefficient ρ_i (in kN/mm) of each layer, of the track structure, is a "spring coefficient" that contributes to the total track stiffness coefficient ρ_{total}. The calculation of stiffness ρ_i and ρ_{total} is used to determine the action/reaction on a sleeper. It is cited that after experimental on-track investigation, the theoretical subsidence is the same with that deriving from the calculation of the track vertical stiffness [EIS 84]. Professor Eisenmann [EIS 80, EIS 88] also ascertains that theoretical calculations performed for the dimensioning of the superstructure correspond to the average value of the measurements. The calculation of ρ_{total} is performed for springs in a parallel arrangement by the following equation [GIA 04]:

$$\frac{1}{\rho_{total}} = \frac{1}{\rho_{rail}} + \frac{1}{\rho_{pad}} + \frac{1}{\rho_{sleeper}} + \frac{1}{\rho_{ballast}} + \frac{1}{\rho_{substructure}} = \sum_{i=1}^{5} \frac{1}{\rho_i} \qquad [36.2]$$

The individual values for the static stiffness coefficients ρ_i, for each layer, as they have derived from measurements on track are as follows [ALI 84, PRU 66, PRU 70]:

– ρ_{rail} ranging from 50,000 to 100,000 kN/mm with an average of 75,000 kN/mm;

– $\rho_{sleeper}$ ranging from 500 to 800 kN/mm for oak wooden sleeper and 12,000 to 15,000 kN/mm with an average of 13,500 kN/mm for a concrete sleeper;

– $\rho_{ballast}$ on the order of 380 kN/mm for ballast 2 years after laying, relatively "fouled" ;

– $\rho_{substructure}$ ranging from (i) 20 to 60 kN/mm for pebbly subgrade, (ii) 80 to 100 kN/mm in the case of well-compacted substructure, (iii) between 86 and 171 kN/mm for NBS of the German railways [EIS94] and (iv) on the order of

250 kN/mm for the case of a ballast bed of small thickness laid on the concrete base of a tunnel or a bridge deck (see [GIA 11]);

– ρ_{pad}, the fastening pad stiffness, plays a key role in the loading of the sleeper, and its value is typically estimated from the load-deflection curve (Figures 36.4 left for the clip and Figure 36.4 right for the pad) that is provided by the manufacturer using a trial and error method [GIA 12a]. For a description of the method of Giannakos [GIA 04, see Giannakos and Loizos [GIA 10b]. Furthermore, the pad stiffness, ρ_{pad}, plays a key role in the loading of the sleeper, and its value is typically estimated from the load–deflection curve that is provided by the manufacturer using the trial and error method [GIA 12c].

In the American literature [KER 02, BAT 99], measurements of the stiffness of track on site can been found. However, since the theoretical analysis should consider the envelope curve for the most adverse conditions, the aforementioned values are more appropriate in order to estimate the "bearing capacity" and the stiffness of the track, according to equation [36.2].

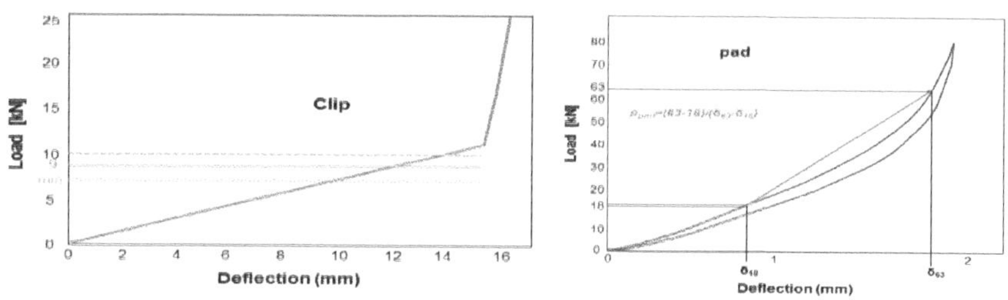

Figure 36.4. *Typical load–deflection curves of fastening clip (left side) and fastening pad (right side)*

It has to be underlined that the fastening and substructure stiffness contributes over 85% to the total static stiffness coefficient of the track and the final values of actions/reactions [GIA 12b]. Professor Eisenmann [EIS 88, EIS 81, EIS 80] also supports that the most determinant parameters, for the definition of the track vertical stiffness, are the quality of substructure and elastic pad, both of which characterize the subsidence (or the stiffness) of a track, that is the distribution of loads between the sleeper, that carries the axle, and the adjacent sleepers. Furthermore, in both cases, for high-speed and heavy haul railways, in order to calculate the acting forces on the superstructure and the sleepers by applying the aforementioned equation in a multi-layered construction with poly-parametrical function, the exact rigidity of the elastic pad of the fastening for each combination of parameters must be determined.

In the case of the resilient fastenings, the most adverse load–deflection curve of the pad is considered because it describes the behavior of the pad during the approach of the wheel. The stiffness of the substructure varies from 40, for pebbly substructure, to 250 kN/mm, for rocky tunnel bottom, or concrete bridge with not enough ballast thickness. Each time this stiffness changes in equation [36.2] above, the "acting" stiffness of the fastening pad also changes [GIA 12a]. The aforementioned calculation of stiffness ρ_{total} is also used in the literature where it is determined that, after experimental *in situ* tests, the theoretical subsidence is the same with that deriving from the calculation of the track's vertical stiffness [EIS 84]. Furthermore, professor Eisenmann [EIS 88, EIS 80] supports that theoretical calculation of the superstructure corresponds to the average of the measurements' results. Since the European Rail Research Institute (ERRI)/ORE [ERR 94a, ERR 94c, ORE 69] results show that the study of the track's construction, on the basis of its vertical stiffness, gives trustworthy results, what should be investigated next is the experimental verification of the outstanding difference in the performance of the fastenings. The total static stiffness coefficient for the case of ballasted track varies from 23.20 to 44.84 kN/mm. On the contrary, the total static stiffness coefficient of the ballastless track equipped with a very resilient fastening obtains values within a very narrow range between 14.83 and 16.90 kN/mm [GIA 12b]. It is important to note that although the slab track is a much more rigid structure than the ballasted track, the total static track stiffness of slab track is 36–62% lower than the ballasted track stiffness. The stiffer the substructure is, the more decisive the contribution of the pad stiffness to the total track stiffness modulation.

36.5. Compatibility of clip and pad

The compatibility of pad and clip is calculated in two phases (see Figure 36.4): in the first phase (fastening's mounting/ tightening) the subsidence (δ_{18}) of the pad is calculated, for the initial clamping force (18 kN approximately), that the two fastenings exert on each rail. In the second phase, the deflection due to the additional pressure exerted on the elastic pad is calculated. This additional pressure is derived at the instance when one axle is situated exactly above the specific sleeper. It can be roughly assumed that the static load, which each semi-sleeper undertakes, is approximately $0.40 \times 112.5 = 45$ kN, given that the rest of the wheel load is distributed to the adjacent semi-sleepers sited on both sides. Figure 36.4 displays the calculation of the reaction coefficient of the elastic pad $\rho_{18\text{–}68}$ (final load = 18 + 45 = 63 kN). In practice, the load undertaken by each sleeper depends on the ρ_{total} of the track at each point. It is easy to accurately calculate it with the method of trial and error, starting from the assumption that 40% of the wheel load exerted on each semi-sleeper is undertaken by the semi-sleeper and after two or three trials, the procedure rapidly converges on the accurate value. When the value ρ of the pad has been calculated, then, from the load–deflection curve of the clip, the decrease in the clamping force can also be calculated. In order to secure the anchorage efficiency (clamping force) of the foot of the rail in each position, the remaining clamping force per fastening (min) should be higher than 8 kN.

36.6. Influence of pad stiffness on the stresses on ballast

The need to use highly resilient fastenings and pads, practically imposes the use of resilient and thick pads, in order to achieve larger deflections and smaller stresses on ballast bed. The increase in the stresses on the ballast bed, due to the use of fastenings with stiff (and thin) pads, in comparison to resilient fastenings with resilient (and thicker) pads (e.g. W14+Zw700), fluctuates from 17.1 to 34.1% [GIA 12a]. These percentages imply an increase in the annual maintenance cost – according to the fourth rule in equation [36.1] – between 88 and 223%. It is obvious from the analysis above that the type of the fastening used – and specifically its resilient (thicker) pad – plays a major role in the magnitude of the stress on the contact (seating) surface between sleeper and ballast bed. Track maintenance and renewal are planned accordingly, always taking into consideration local conditions, based on resultant of control data from measuring systems, visual observation and economic data. Each track, as in every construction, has a predetermined lifecycle during which maintenance works are necessary for the provision of the basic, minimum standards of quality and safety as mentioned above. For conventional superstructure consisting of rail, fastenings, sleepers, and ballast, an optimum lifecycle exists from an economic point of view based on the traffic daily/annual tonnage.

36.7. Requirements for the fastenings and pads due to their role in the track panel

The disadvantage of non-elastic fastenings is that the quality of their functions deteriorates rapidly under the impact of rolling loads and the passing of time due to the fact that the induced deformations are permanent. The fact that the absorption of energy is cumulative implies that the deformations created are of progressive nature, which is the basic cause of permanent deformations and unstable equilibrium on the track. Elastic fastenings, on the contrary, are capable of being deformed and reverting to their initial condition after the deformation unless their elastic characteristics have been deteriorated. The quality of the latter depends of course on how well these fastenings have been designed and produced. Due to their elasticity, these fastenings – compared to rigid fastenings – maintain the quality of the track and the ease of its movement over a longer period of time. In order to analyze the various forms of the fastenings' functions, we should take into consideration the nature of the loads imposed as well as their components.

There are fastenings where the clamping force on the rail foot is achieved through friction and not through tightening of a screw. The clip is in direct contact and exerts the force directly on the rail foot, thus bringing about wear of the metal parts and loosening of the clamping force. If there is no possibility for periodical correction of the clamping force (through a bolt with thread), gradual and permanent decrease in the clamping force actually exerted will ensue. There are also fastenings, where the pressure on the rail foot is achieved through the interposition of a pad (between the clip and the rail foot) and in this case there is no wear of the metal parts. The pad can be replaced more easily. Fastenings, where the clamping force on the rail foot is constant and does not fluctuate (by tightening) during their use, because there is no bolt with thread permitting it, present a loosening with time.

The lifecycle of the track is dependent upon the fatigue of the repetitive loading and it is highly influenced by the action and stress reduction [GIA 12a]. Professor Eisenmann [EIS 88] supported that for the W1 fastenings of DB combined with pad Zw687, the value ρ_{pad} = 450 kN/mm is valid and he cites that the elastic pad should reach ρ_{pad} = 100 ÷ 150 kN/mm. The necessity for technological development and production of fastenings with higher elasticity (ρ_{pad} = 100 kN/mm) means (1) production of an elastic pad with high elasticity, which will last longer; (2) production of a doubly elastic fastening clip, of relevant elasticity, compatible with the pad, which means that, during maximum subsidence of the pad, the minimum clamping force on the foot of the rail will be secured, approximately 800 ÷ 900 kg for each fastening, that is per side of the foot of the rail. This means that if the fastenings are correctly tightened, then the elastic pad has a permanent minimum compression force 1600 ÷ 1800 kg; and (3) compatibility of clip pad.

The German Railways (technical specifications TL 918 127) specify compatibility as a function of the elastic pad reaction coefficient (ρ or c) and response of the fastening to a dynamic sinusoidal load with a determined amplitude of dynamic oscillation. The compatibility of both the elastic pad and the clip is ascertained (certified) by the load–deflection curves of these two individual components of the fastening. The use of the resilient fastenings instead of the rigid fastenings leads to a very impressive reduction of the annual maintenance cost of the railway track infrastructure. The technological progress in pad production led to a new generation of very resilient products of better quality. According to the Greek and German regulations (DB-TL 918 235, 1994), the pads should provide a lifecycle of 20 years. The corresponding very resilient fastening clip should provide a lifecycle of 30 years. Moreover, according to research performed by ERRI (U.I.C.), the W clip (e.g. the clip of W14) stops at a deflection of 2 mm rail tilt [ERR 94a, ERR 94b) and presents sufficient strength. The development of "channeled" or "studded" pads permitted (technologically) the increase in elasticity of a pad (production of very resilient pads) at the order of 45–50 kN/mm because of the "air circulation", which facilitates the heat removal and the (probable) water sewerage [GIA 04]. This should imply an avoidance of the pad's degradation/ blowing out and the railseat abrasion. In Europe (with high values of the dynamic component of the acting load), no railseat abrasion has been reported in the cases of very resilient rail pads [GIA 11].

The lifecycles of the pads and the clips of the fastenings are dependent on the actions exerted on each sleeper per passing axle and the number of "axle-passes". Since the actions are of comparable magnitude, the tests for high speed are enough for heavy haul as well. Harisson and Ahlbeck [HAR 87] verified the much better behavior on track and in laboratory of the resilient versus the stiff pads. According to the Greek and German regulations (DB-TL 918 235, 1994), the pads should provide a lifecycle of 20 years (or 400 million tonnes). The corresponding very resilient fastening clip should provide a lifecycle of 30 years (or an estimation of 600 million tonnes).

36.8. Secondary stiffness of the fastening clip and behavior of the track panel

The necessity for adoption of the resilient fastenings – that is of very resilient thick pads – exhibits problems of rail rotation (tilting of the rail) and probable "gauge widening" or "overturn of the rail" with dangerous consequences. The thick pad (for achievement high resilience) has as a disadvantage the great angle φ of rail's rotation, a fact that could lead to the rail's overturning. In Figure 36.5 (left side), the situation of rail rotation is depicted.

This takes place because – under high loads – rotation causes the rail foot to lift on the side nearest the load and this exceeds the preloading imposed when mounting the fastening and provoking a partial lifting of the rail foot from the pad. This fact reduces the effective pad width and consequently the pad stiffness with influences on the stresses, the vertical stiffness of the track and the rotation [ERR 94c).

Thus, a secondary stiffness of the clip is required, a characteristic demanded theoretically for the clip in order to reduce or nullify these phenomena, as depicted in the load – deflection curves of three fastening clips in Figure 36.5 (right side). This characteristic is required more strictly for the tight curves specifically, where the horizontal forces (centrifugal) are increased.

The lifting of the pad is more noticeable when the inclination with which the load is applied is higher. Simplified calculation shows that the center of rotation of the rail foot moves closer to the center of the rail foot as the line of action of the force moves away from the center of the rail foot (eccentricity effect). This means that the pad rises further for a given compression or rotation. Increasing pad stiffness causes a similar smaller geometrical effect and reduces the effective preloading, leading to a greater degree of lift [ERR 94c). In the case of W14, the steep ascent assumes a determinative role only when the rail foot is twisted in such a way as to come into contact with the central loop of the clip. This entails a displacement of up to 2 mm. This does not generally arise since the twist factor is smaller than this and contact with the central loop is (almost) impossible without twisting given the tolerances in question. This confirms that the secondary stiffness of the clip Skl-14, correctly mounted, does not permit an increase in the gauge more than 2 mm [ERR 94a]. The angle (φ in Figure 36.5, left side) of rotation (tilting) for the rail foot is obtained depending on whether the calculation procedure is affected by the distortion of the rail web or the rail foot. Of these two factors, the distortion of the rail foot is negligible, particularly when elastic pads are used. Taking into account that the tests of the ERRI were performed with rail UIC60, we conclude that the distortion of the web and the rail foot will be more serious with a lighter rail. As far as the tilting of the rail using soft or stiff pad the ascertainment is that the rotations (tilting) of the rail in the fastening is on average greater with the flexible pad Zw 700 than it is with the Zw 687a pad (standard design). The relatively small difference, however, does not raise any doubts concerning safety. The comparison of the results obtained from measurements on the track and from measurements on the track panel in the laboratory shows the values measured on the track to be considerably lower than those measured on the track panel in the laboratory for comparable loads. This is a feature of rail fastening W with the rail pads Zw 687a and Zw 700. This is attributable to the greatly increased coefficients of torsion on the track itself, particularly on the outer rail [ERR 94d].

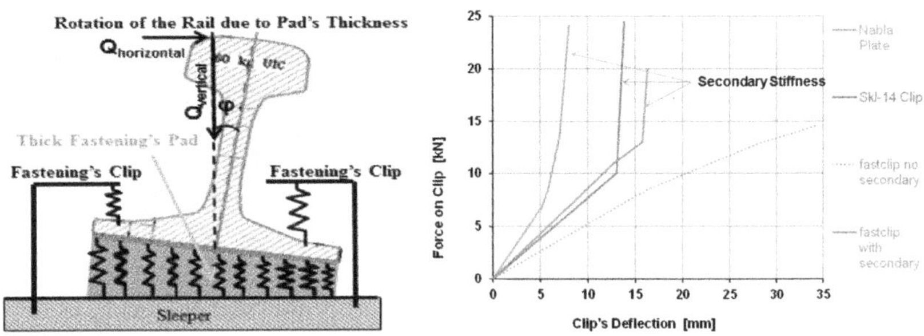

Figure 36.5. *(Left side) Rotation φ of the rail due to pad thickness. (Right side) Load–deflection curves of two clips and one plate [fastclip with no secondary stiffness, Skl 14, Nabla]. The diagram was designed by the author, based on [ERR 94b]. The diagram of the load–deflection curve, of the fastclip with secondary stiffness, was designed by the author, based on the booklet "A new Generation Rail Fastening". For a color version of this figure, see www.iste.co.uk/jacob/materials.zip*

In Figure 36.5 (right side), the load–deflection curves of two clips and one plate (fastclip with no secondary stiffness, Skl 14, Nabla) are depicted. The diagram was designed by the author, based on ERRI D170/DT 283 [ERR 94b]. It should be underlined that one of the tested clips of fastenings does not present secondary stiffness. In this figure (right side), the load–deflection curve of the same clip of fastening (fastclip), but with secondary stiffness, is displayed. The diagram was designed by the author, based on the booklet "A new Generation Rail Fastening" (more details and a similar curve in [RP 95]). The secondary stiffness is due to the use of a shoulder, with additional metallic elements in comparison to the normal one. The results of this investigation (plus the results of the parallel investigations for the acting loads, ballast fouling, cracking of concrete sleepers, etc.) supported the decisions of the Greek Railways for the high-speed lines in Greece (since the 1990s).

36.9. Conclusions

The requirements for a modern railway track fastening should be as follows: (1) low static stiffness coefficient ρ_{static}<100 kN/mm for the fastening as derived from the load–deflection curve of the pad, (2) compatibility of the clip and the pad of the fastening as it is derived from the combination and comparison of the load–deflection curves of the clip and the pad (toe-load in any case >8 kN), (3) secondary stiffness in order to decrease the "tilt of the rail" and keep it under a maximum limit of approximately 2 mm.

36.10. Bibliography

[ALI 84] ALIAS J., *La voie ferrée*, Eyrolles, Paris, 1984.

[BAT 99] BATHURST L., KERR A., "An improved analysis for the determination of required ballast depth", *Proceedings of AREMA Conference*, 1999.

[EIS 04] EISENMANN J., "Die schiene als tragbalken", *Eisenbahningenieur*, vol. 55, pp. 22–25, May 2004.

[EIS 97] EISENMANN J., RUMP R., "Ein Schotteroberbau für hohe Geschwindigkeiten", *ETR*, vol. 46, H.3, pp. 99–108, March 1997.

[EIS 94] EISENMANN J., LEYKAUF G., MATTNER L., "Vorsläge zur Erhöhung der Oberbauelastizität", *ETR* 43, H 7/8, July August, pp. 491–494, 1994.

[EIS 88] EISENMANN J., "Schotteroberbau – Moglichkeiten und Perspektiven fur die Moderne Bahn", *Die Naturstein-Industrie*, Heft 3, Isernhagen, Germany, pp. 6–11, 1988.

[EIS 84] EISENEMANN J., MATTNER L., "Auswirkung der Oberbaukonstruction auf die Schotter – und Untergrundbeanspruchung", *Eisenbahningenieur*, vol. 35 no. 3, 1984.

[EIS 81] EISENMANN J., "Verhaltensfunktion des Schotters, Folgerungen fur hohe Fahrgeschwindingkeiten", *Der Eisenbahningenieur*, Darmstadt, March 1981.

[EIS 80] EISENMANN J., "Verjormungsverhalten der Schiene.Auswirkungen auf die Oberbaubeanspruchung, Lagestabilitat und Storungsfunktion", *ZEV-Glasers Annalen*, Berlin, February 1980.

[ERR 94a] ERRI, D 170/DT 282, "Calculation of twist and displacements in UIC60 rail using rail fastening system W and rail pads Zw687a and Zw700", *UIC – International Union of Railways-ERRI*, Utrecht, September 1994.

[ERR 94b] ERRI, D 170/DT 283, "Response of the railway track under external loading", *UIC – International Union of Railways-ERRI,* Utrecht, September, 1994.

[ERR 94c] ERRI, D 170/DT 302, "Stresses on the fastening system under the action of wheel loads", *UIC - International Union of Railways-ERRI,* Utrecht, September, 1994.

[ERR 94d] ERRI, D 170/DT 304, "Rotation (Tilting) of the rail in the w fastening when using pads of different flexibility", *UIC – International Union of Railways-ERRI*, Utrecht, September, 1994.

[GIA 12a] GIANNAKOS K., "Comparison of magnitude of actions on track in high-speed and heavy-haul railroads – influence of fastenings", *Transport Research Record – TRR*, no. 2289, pp. 70–77, 2012.

[GIA 12b] GIANNAKOS K., "Modern railway infrastructure: resilient fastenings improve track's life-cycle", *Proceedings of TRA2012*, Athens, 23-26 April, 2012.

[GIA 12c] GIANNAKOS K., TSOUKANTAS S., "transition zone between ballastless and ballasted track: influence of changing stiffness on acting forces", *Proceedings of TRA2012*, Athens, 23-26 April, 2012.

[GIA 12d] GIANNAKOS K., "Heavy haul railway track maintenance and use of resilient versus stiff fastenings", *Transport Research Record – TRR,* 2261, pp. 155–162, 2012.

[GIA 10a] GIANNAKOS K., "Loads on track, ballast fouling and life-cycle under dynamic loading in railways", *International Journal of Transportation Engineering – ASCE*, vol. 136, no. 12, pp. 1075–1084, 2010.

[GIA 10b] GIANNAKOS K., LOIZOS A., "Evaluation of actions on concrete sleepers as design loads–Influence of fastenings", *International Journal of Pavement Engineering (IJPE),* vol. 11, no. 3, pp. 197–213, June 2010.

[GIA 04] GIANNAKOS K., "Actions on the railway track", *Papazissis publ.*, Athens, Greece, available at www.papazisi.gr, 2004.

[HAR 87] HARISSON H.D., AHLBECK D.R., "Railroad track structure performance under wheel impact loading", *Transport Research Record – TRR,* 1131, pp. 239–264, 1987.

[KER 02] KERR A., "The determination of the track modulus k for the standard track analysis", *Proceedings of AREMA Conference*, 2002.

[ORE 83] ORE D117, Rps 1 – 27, "Optimum adaptation of the conventional track to future traffic", *UIC - International Union of Railways–ORE,* Utrecht, Belgium, 1971–1983.

[ORE 69] ORE, D71, Rp9, "Sollicitations de la voie du ballast et de la plateforme sous l'action des charges roulantes", *UIC – International Union of Railways-ORE*, Utrecht, October, 1969.

[PRU 76] PRUD' HOMME A., ERIAU J., "Les Nouvelles Traverses en béton de la S.N.C.F.", *Revue Générale des Chemins de Fer*, February, 1976.

[PRU 66] PRUD' HOMME A., "Sollicitations statiques et dynamiques de la voie", *S.N.C.F. – Direction des Installations Fixes*, 3/1966, 1966.

[RP 95] RP. No 1366, "Pandrol fastclip – laboratory and track testing of a captive switch on/ switch off rail fastener", *Pandrol Rail Fastenings Limited*, Surrey, U.K, 1995.

[WEA 90] WEAVER W.J.R., TIMOSHENKO S.P., YOUNG D.H., *Vibration Problems in Engineering*, 5th edition, John Wiley & Sons, 1990.

[TL 98] TL 918 127, Technische Lieferbedingungen Spannklemmen und Spannagel-TL 918 127, Deutsch Bahn (DB), Germany, 1998.

[TL 94] TL 918 235, Elastische Zwischenlangen, Deutsch Bahn (DB), Germany, February, 1994.

[WIN 67] WINKLER E., "Die lehre von der elastizität und festigkeit (the theory of elasticity and stiffness)", *H. Dominicus*, Prague, 1967.

[ZIM 41] ZIMMERMANN H., "Die Berechnung des Eisenbahnoberbaues", *Verlag von Wilhelm Ernst & Sohn*, Berlin, 1941.

37

A New Asset Management Approach for Inland Waterways

Inland waterways in Europe face a resource patchwork, low availability and effectiveness of investments leading to a declining importance compared to other modes of transport. To overcome these setbacks, the presented waterway asset management approach aims at an increased availability of fairway widths and depths in days per year. Based on periodic riverbed surveys, current water levels and discharge, the impact of maintenance and river engineering works on the availability of fairway widths and depths in days per year can be analyzed. Any resulting increase in availability on a transport route leads to a decrease in transport costs that are considered as a benefit of the implemented measures. Innovative alert systems based on an empirically derived behavior of critical bottlenecks allow preventive measures and lead to real-time availability information for the transport industry. The approach enables an optimization regarding individual measures on river sections as well as of investment strategies for constrained budgets, recommended fairway conditions or total costs of waterway agencies and the transport industry. The presented asset management approach is currently being implemented on the Austrian stretch of the river Danube based on a new software tool and may be extended to the entire waterway in the future.

37.1. Overview waterway asset management

Waterway asset management is a holistic multidisciplinary approach for the development, maintenance, rehabilitation and replacement of waterway assets based on a comprehensive lifecycle costing approach. According to waterway authorities [VIA 13a], river navigation is only shut down in times of extreme weather conditions (floods, extensive ice formation) but usually not at low-water conditions. Based on this definition, the average availability of the Austrian section of the Danube

---

Chapter written by Markus HOFFMANN, Katrin HASELBAUER, Alexander HABERL, Ronald BLAB, Markus SIMONER, Klaus DIEPLINGER and Thomas HARTL.

waterway for the period 1997–2012 was calculated with 97.9%. Passenger transport in the tourist season with a vessel draught usually not exceeding 1.5 m faces almost no restrictions on the Danube even at lowwater conditions. In comparison, the transport of particular cargo types will only be competitive if the provided fairway parameters are sufficient. In order to ensure continuous and satisfactory fairway conditions, there are a number of recommendations and international agreements (e.g. AGN, DC1998, DC2013) aiming at common minimum fairway parameters (width, depth, days per year). However, due to political, technical, environmental and economic reasons, necessary fairway maintenance and river engineering works on the Danube are not implemented in a sufficient way resulting in unsatisfactory fairway conditions. Furthermore, maintenance and river engineering works are currently planned on a case-by-case basis with limited continuous coordination and optimization between riparian countries.

To overcome these setbacks, the presented asset management approach (Figure 37.1) is based on the availability of fairway widths and depths in days per year. With a three-dimensional data model of the river including locks, ports and the navigational fairway, the impact of maintenance measures on the availability of the waterway can be modeled. An increased availability will in turn lead to a higher utilization and lowered costs of inland navigation. Based on a fleet model and an availability-based transport cost model, the resulting transport cost savings can be calculated for different investment strategies. Based on this approach, cost-efficient measures can be found not only for individual critical sections but also for the entire Danube. Furthermore, it is possible to optimize with regard to the recommended fairway parameters, a constrained budget or minimal total costs of inland navigation. The following sections of this chapter will highlight certain aspects of this asset management approach and the developed optimization process.

37.2. Maintenance measures and fairway availability

37.2.1. *Fairway parameters and availability*

For navigation companies, the availability of minimum fairway widths and depths is crucial both for planning individual transport procedures and for being competitive throughout the year. In order to calculate the availability for a cross-section profile, a river section or for entire transport routes the water level based on data from automatic gauges and periodic riverbed surveying are needed. The physical availability for a river section on a specific day results from the non-intersection of a combination of width and depth of the fairway path with the riverbed (Figure 37.2(a)). If this procedure is repeated for all days of a period, the availability is defined as the number of days with non-intersection divided by the total number of days in this period. Repeating this calculation for any combination of fairway width and depth results in a decreasing availability with increasing fairway dimensions (convex falling performance surface). Thus, any recommendation may be described as a single point that may be below or above the actual availability surface. Additional measures (e.g. dredging or relocation of fairway) have to be implemented if this point is above the availability surface (Figure 37.2(b)).

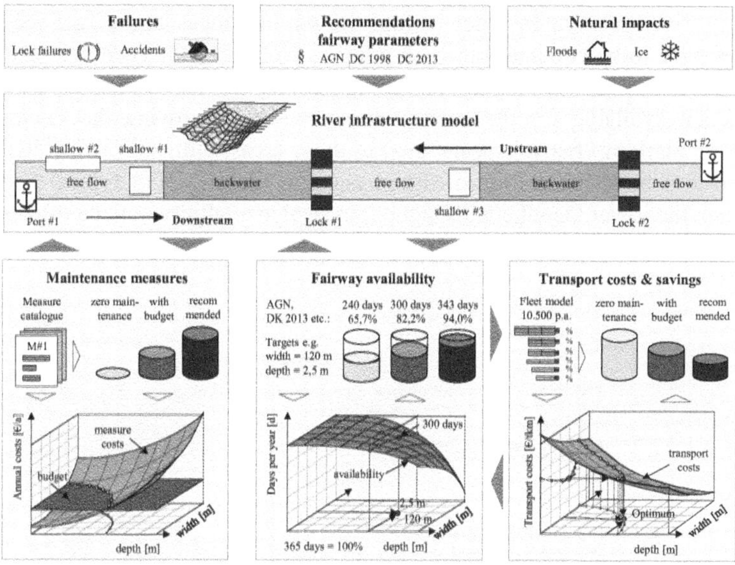

Figure 37.1. *Asset management approach based on fairway availability with measure costs and transport cost savings*

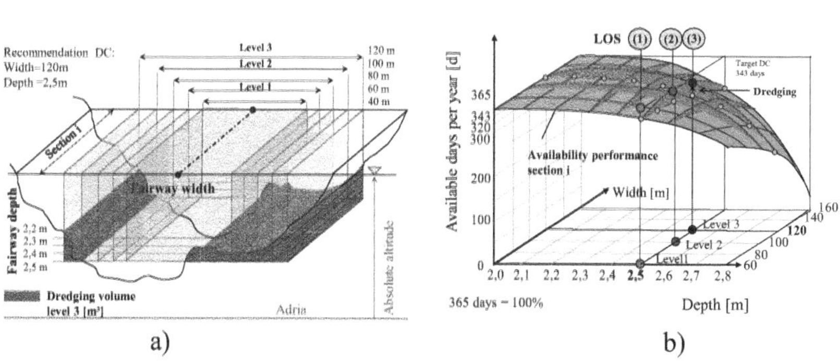

Figure 37.2. *a) Classes of fairway width and depth of a river section i with the actual riverbed condition on daily basis; b) availability performance of the river section i for 1 year with deviation from different levels of service (LOSs), e.g. regarding width (e.g. LOS 3 = DC1998 → width = 120 m, depth =2.5 m, 343 days)*

37.2.2. *Measure overview with decision tree and impact*

In order to increase fairway availability, waterway agencies may choose between various possible measures with different costs, impact on availability, realization time, duration of impact, resulting user costs and environmental impacts. Figure 37.3 shows a decision tree of possible operation, maintenance and engineering measures as well as the availability performance for a fixed fairway depth of a river section. In order to find the optimal measure for one section, all measures have to be compared to each other as well as to the current situation that may be described as “status quo” or “doing nothing”. If the recommended fairway depth is not available for the entire fairway width, narrowing the fairway with appropriate marking (buoys, spears, etc.) to those areas with sufficient water depths may lead to a better utilization of the physical availability. For typical wide river sections on the lower Danube with higher physical availability outside of the current fairway, shifting of the course of the fairway may also be an option.

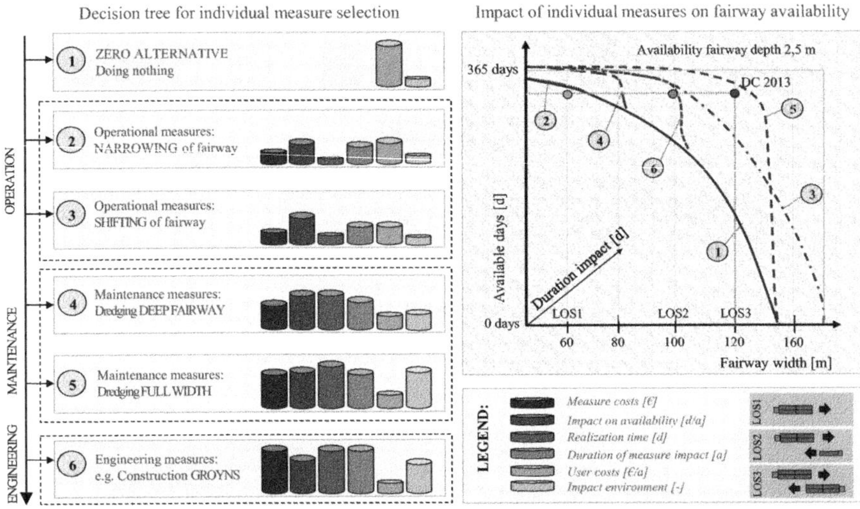

Figure 37.3. *Overview decision tree for a selection of individual measures and their resulting impact on fairway availability*

For river sections without sufficient widths and depths, only physical measures may lead to an increase in fairway availability. The fastest and least costly measure would be to dredge a “deep fairway channel” in order to achieve the necessary depth on a minimum necessary width (e.g. for one-way traffic). On river sections with very high traffic density and no budgetary or environmental restrictions, dredging to the full fairway width may be an alternative. However, on river stretches with very dynamic morphology, the duration of dredging impact may be insufficient leading to the question of more lasting measures. Such measures are defined as river engineering measures and include groynes and training walls for low water regulation. Typically, planning and implementation of this kind of measure takes longer and they are costlier as well. For all measures, a careful assessment of the individual situation is always necessary due to different types of river morphology and resulting deviations in costs and impact. For successful planning and implementation, periodic riverbed surveying is mandatory as well as an assessment of actual impacts and costs of already implemented measures.

37.2.3. *Measure costs, impact and duration*

For a lifecycle cost analysis in a waterway asset management system, the comparison of possible measures is always based on costs and duration of impact. Therefore, it is necessary to derive cost functions from previous examples depending on measure extent in order to account for the economy of scale. The duration of measure impact is defined as time with increased availability compared to "doing nothing". For dredging as main measure on the river Danube in Austria, Figure 37.4(a) provides the information on actual unit costs. Figure 37.4(b) explains the duration of dredging impact based on the backfilling rate of dredged volume. The priority of further measures may then be determined by the progressivity of backfilling behavior and percentage of backfilling.

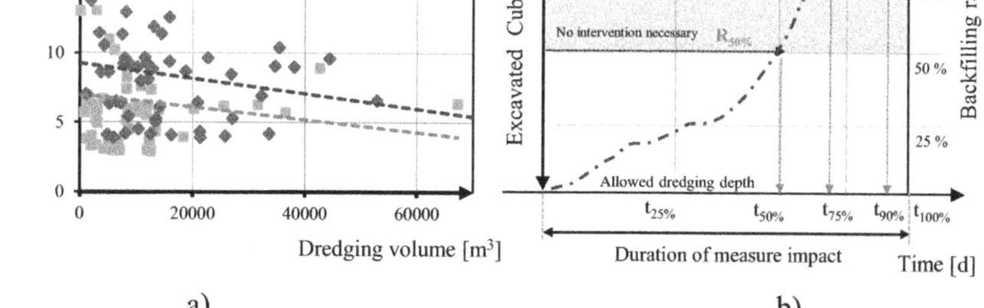

a) b)

Figure 37.4. *a) Costs per unit of fine sediment and gravel dredging depending on measure extent (economy of scale); b) duration of dredging measure impact with specific backfilling rate being related to various parameters (e.g. amount of discharge)*

37.2.4. *Optimization for continuous fairway parameters*

Finding the optimal measure for a specific river section is not enough in order to provide continuous fairway parameters on a certain level of service (LOS) for inland navigation. For a transport route between two ports with several critical sections in between, the shallowest section limits the maximum loaded draught of a vessel, whereas restrictions in fairway widths may lead to additional waiting times in case of critical encounters. Figure 37.5 provides a schematic overview of such a transport route with the availability performance of widths and depths of all critical or shallow sections. In order to achieve continuous fairway parameters for any given LOS, the availability performance has to be increased step by step, starting with the most critical shallow section. For a serial system, the effectiveness of measures on availability performance is therefore limited depending on the condition of the next most critical section. Further expenditures toward improving the availability on the first critical section are thus a waste of budget if the conditions on the next critical sections are not improved as well. For a manual measure optimization, the availability curves may be shifted outward until the budget is spent. Subsequently, the necessary extent and costs of all measures on a transport route will also be different depending on the targeted condition of the fairway or LOS. In Austria, these LOSs are based on a constant minimum fairway depth of 2.5 m at low navigable water level and a fairway width of 40–80 m (LOS1), 80–120 m (LOS2) and 100–160 m (LOS3), depending on bend radii and describing critical encounters.

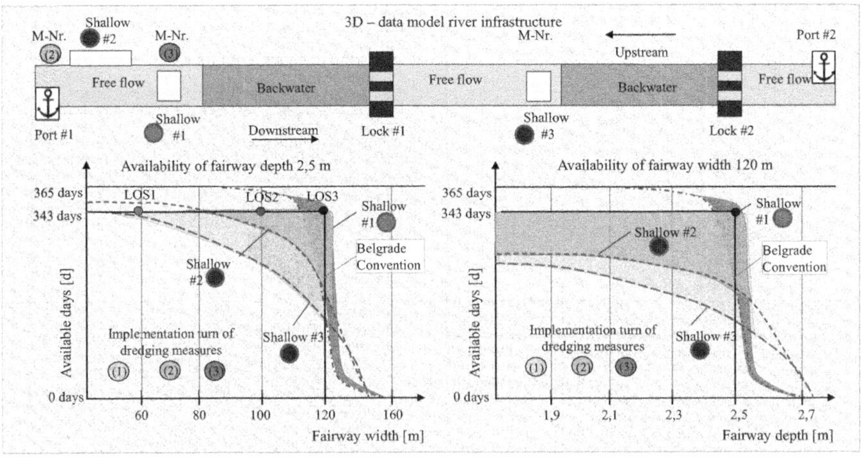

Figure 37.5. *Optimization for continuous fairway parameters and resulting increase in availability for an entire transport route*

37.2.5. *Measure program, priorities, budget and market capacity*

As a result of this optimization process, all necessary measures aimed at achieving a certain LOS may be condensed into a measure program containing priority, measure extent and costs as well as the time for implementation. Figure 37.6 shows the concept of such a program for different LOSs with a main emphasis on, but not limited to, dredging measures. On the most critical sections, regarding availability, certain measures (e.g. dredging) will be necessary even for lower LOSs (e.g. LOS1). With a higher targeted LOS, the number of critical sections with necessary measures will increase. The same holds true for the necessary measure extent as well as the resulting costs and time for implementation. The priority between measures to achieve a certain LOS is based on the critical behavior of individual sections (alert system). Thus, possible negative impacts on inland navigation can be minimized. All measures to achieve a certain LOS result in a total measure extent, costs and time. The implementation of the resulting measure program may be limited by several factors. Usually, the time frame for possible interventions in the riverbed is restricted to certain periods due to environmental reasons (e.g. spawning season of fish). Further restrictions and shifts in priorities may be applied due to periods with low water levels. For dredging as the main measure in Austria, the possible dredging volume may also be limited due to the currently available market capacity. Even with a sufficient budget to hand, it would not be possible to implement all measures in time to achieve a certain level of fairway availability and minimize negative impacts on inland navigation.

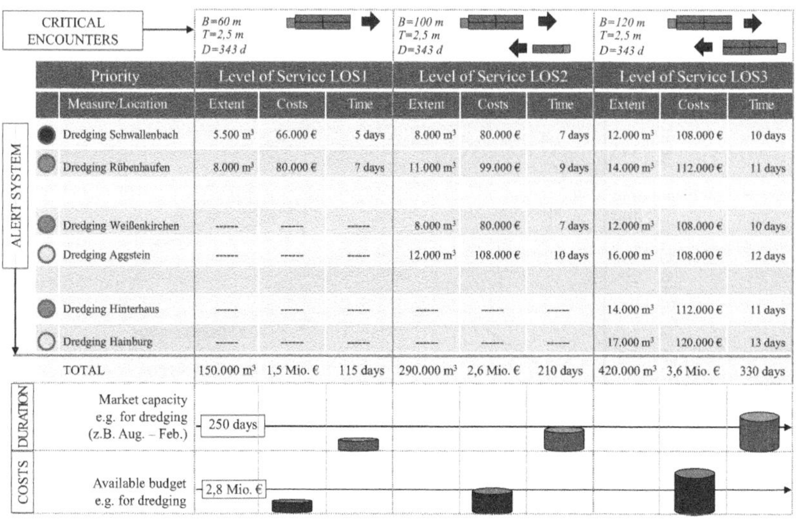

Figure 37.6. *Resulting measure program for dredging measures with measure extent, measure costs and time for implementation depending on targeted level of service and priority (e.g. due to critical condition or development)*

37.3. Fairway availability and transport costs

37.3.1. *Transport development, vessel types and fairway utilization*

The actual utilization of the fairway is always lower than its physical availability and depends on traffic volume, fleet composition and critical encounters (Figure 37.7). Further limiting factors may be the predictability of water levels especially for long transport routes as well as the perceived reliability of provided information on fairway parameters or current situation on the transport market. According to the annual reports of the Austrian waterway management company viadonau, the transport capacity utilization average is 12–15%. In goods transport, the traffic volume is aligned with the actual water depths limiting loading volumes leading to higher transport costs. The average load factor for cargo vessels on the upper Danube was around 60–65%. On the lower Danube, the average loading factors are between 50 and 55%. With a typical vessel draught usually not exceeding 1.5 m, passenger transport faces almost no restrictions but is mainly concentrated between April and November. For the utilization of fairway availability, an analysis of the traffic volume and vessel fleet may provide further insights. The total number of ships passing the Altenwörth power plant in 2012 upstream or downstream amounted to 10,700 with 35% being passenger vessels and 65% cargo vessels. The majority of cargo vessels are individual vessels followed by two-unit pushed convoys. With a share of 4.3% of the cargo vessel fleet, the critical encounter probability of pushers with four lighters (LOS 3) in one of the short critical sections is rather low. Further analysis of transponder data proves the low utilization of fairway width, which is therefore not a critical issue on the River Danube.

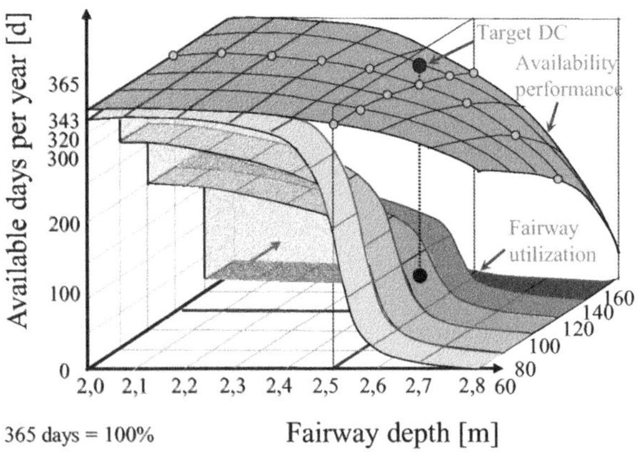

Figure 37.7. *Utilization of physical fairway availability based on traffic volume, fleet composition and critical encounters*

37.3.2. *Impact of fairway depth on utilization of loading capacity and transport costs*

The fairway depth needed for the trip of an individual vessel consists of a static draught, the dynamic squat and an underkeel clearance and must be lower than the actual provided fairway depth. The static loaded draught is a result of the utilization of possible transport volume and depends on the vessel type. Figure 37.8(a) provides an overview of such calibration curves for the cargo fleet on the River Danube based on data from different sources (e.g. [SIM 04]). At a typical load factor of 60%, the most common single vessel (type Johann Welker) provides a static loaded draught of around 2.2–2.3 m. This is much lower compared to large motor cargo vessels or pusher and lighter combinations at the same utilization, which are more suitable for the transport of large quantities of goods at high water levels. The squat depends, among others, on travel speed, the cross-section profile of the waterway and water depths. Figure 37.8(b) provides an overview of typical squat values for sections with a shallow riverbed according to Simoner [SIM 04]. In summary, the provided fairway depth directly relates to a specific utilization of cargo vessels and is the main factor for the resulting unit costs per ton kilometer.

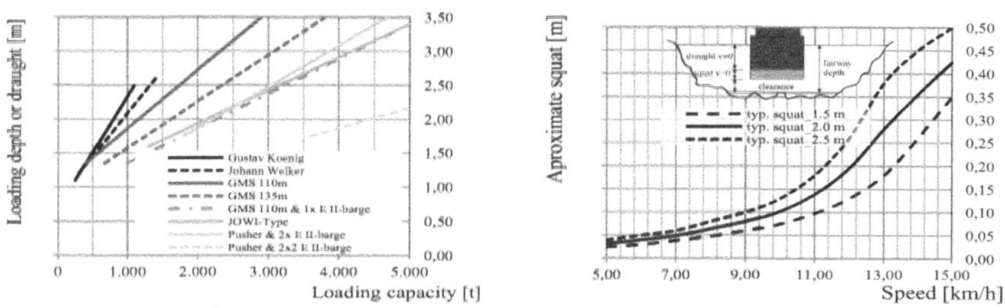

a) b)

Figure 37.8. *a) Static draught/loaded depth and loading capacity of Danube fleet; b) dynamic squat depending on static draught and speed for low water levels and main vessel types of the River Danube fleet*

37.3.3. *Utilization-based transport costs and market competition*

In addition to vessel utilization, the reliability, availability and resulting transport costs are of crucial importance regarding competitiveness of different modes of transport on the market. Depending on the type and amount of goods, transport relations, transport distance and possible utilization, the choice for a mode of transport or intermodal transport chain will be different. The presented transport cost comparison is based on several sources (e.g. [PLA 07, SCH 10]) and additional research. Figure 37.9 shows transport unit costs for inland navigation including pre- and end haulage within a radius of 50 km for the most common vessel types. The unit costs show a decreasing behavior with increasing transport distance and vessel utilization. Due to fuel costs, transport upstream is significantly more expensive than transport downstream. The high amount of fixed handling costs but lower total transport costs make inland navigation attractive especially for bulky goods on long transport distances exceeding 500 km. First results of transport cost analysis indicate that an average vessel load factor of at least 55–60% is necessary to stay in the market.

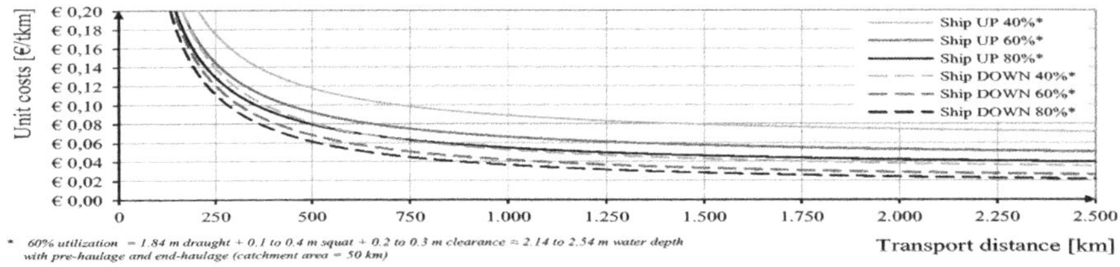

Figure 37.9. *Transport distancerelated unit costs of inland navigation on the Danube (e.g. for main Johann Welker type vessel)*

Per centimeter of utilized additional draught, the transport capacity increases sufficiently (e.g. 7.8 t for Johann Welker type vessels). Thus, possible savings of an improved availability of fairway depth can also be estimated based on this transport cost model. The total savings are a result of the utilization of additional fairway depths (and widths) for the entire transport volume. Any utilization of provided fairway widths and depths leads to total transport costs resulting in a concave falling transport cost surface with increasing fairway dimensions. The possible transport cost savings for a measure program may be obtained as the difference between transport costs and measures compared to "doing nothing". If external costs are included, possible economic savings in inland navigation due to improved fairway conditions will be a few times higher as inland navigation shows the lowest negative impact on the environment compared to other modes (e.g. [PLA 07, ECC 12]).

37.3.4. *Transport costs and fairway availability optimization*

In Figure 37.10, all necessary input parameters for an optimization of investment strategies are compiled. Any increase in availability will lead to certain total measure costs resulting in a concave increasing cost surface with increasing fairway dimensions. The given budget defines a new increased availability surface. If the targeted LOS (LOS1, LOS2 and LOS3) is above this budget-related availability surface, it cannot be achieved without further funding. However, this does not necessarily lead to optimal conditions for inland navigation due to the fact that it is very unlikely that these predefined conditions will be costefficient with respect to transport costs.

In general, measures may be optimized regarding different fairway parameters leading to specific combinations of continuous fairway widths and depths in days per year that may be achieved within the same budget. The resulting combinations may be also modeled as the intersection of the horizontal annual budget surface with the increasing measure cost surface for achieving increasing availability of fairway depths and widths. If this intersection curve is projected on the resulting transport costs surface, the optimal combination of fairway width and depth with minimal transport costs within the given budget can be found (Figure 37.10). For an overall optimum of measure and transport costs, the annual measure costs and availability-based annual transport costs have to be added. The optimal fairway parameters are then at the lowest point of the resulting total annual cost surface.

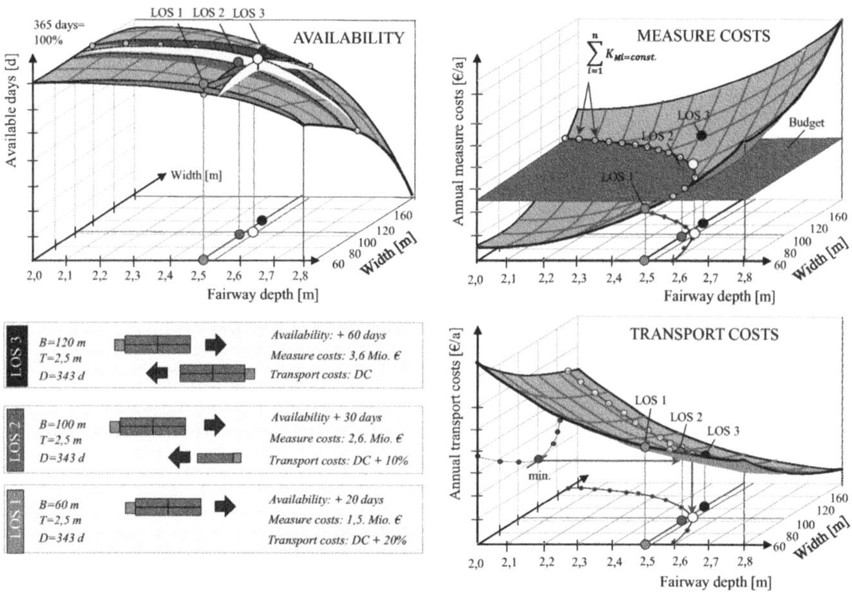

Figure 37.10. *Optimization of fairway parameters for any given level of service with resulting total measure costs and their impact on increasing fairway availability and decreasing transport costs*

37.4. Pilot implementation and first results

The presented asset management approach is currently being implemented on the Austrian stretch of the Danube based on a developed software tool. One goal of this software tool is to create a database and visualization of all relevant waterway infrastructure properties. Furthermore, it will provide transparent process management, benchmarking and documentation tools for all waterway relevant activities.

With this new software tool, it is already possible to calculate fairway availability for any river cross-sectional profile, river section or transport route on a daily basis and given time frame. Figure 37.11 shows the availability of the ford Weißenkirchen for 2012 as well as the typical low water months (e.g. November) and high water months (e.g. May) for 10 years in the period 2003–2012. If the resulting fairway availability surface for 1 year is above the gray cube (DC2013 = LOS3), the implemented measures on this ford have been sufficient.

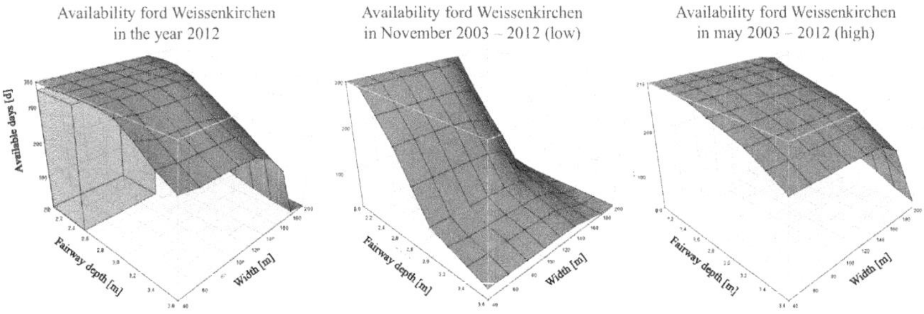

Figure 37.11. *WAMS availability calculation for a shallow section on the River Danube in Austria for an entire year as well as typical low water periods and high water periods*

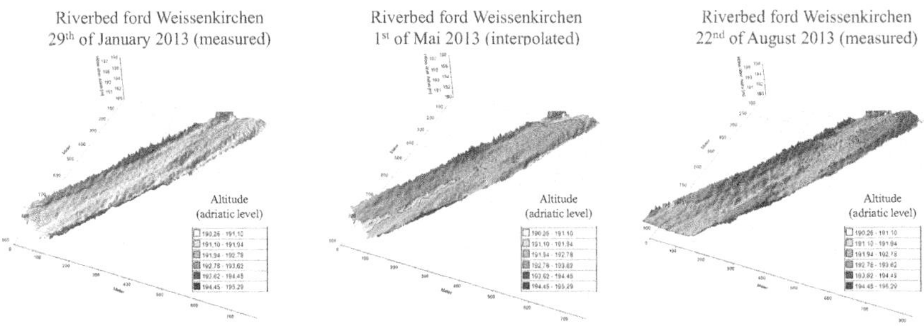

Figure 37.12. *Development of absolute riverbed altitude for a shallow section on the River Danube in Austria with a substantial sedimentation process especially during the flood period in June 2013. For a color version of this figure, see www.iste.co.uk/jacob/materials.zip*

Further functions of the software tool allow for a calculation and visualization of riverbed altitude, water depths as well as sedimentation and erosion processes on a daily basis as well. All information can be processed in real time allowing an animation of all processes over time. Figure 37.12 gives an overview of the development of riverbed altitude in the ford Weißenkirchen between January and August 2013. The sedimentation process is clearly visible especially at the end of the ford and may also be related to the flood of June 2013. The difference map in Figure 37.13 between two respective time frames shows both the impact of an already implemented dredging measure as well as the absolute amount of sedimentation in 2013. With the ability of the developed software tool to interpolate between survey data and calculate changes in volume, it will be possible to analyze the relation between sedimentation and floods as well as the impact duration of implemented measures.

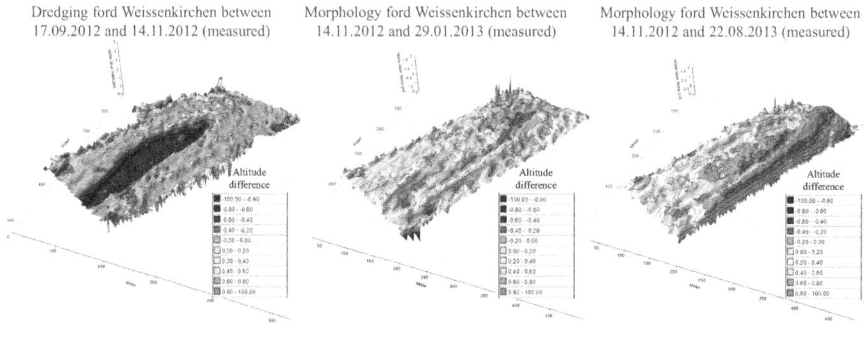

Figure 37.13. *Impact of dredging measure in a shallow section on the river Danube in Austria in October 2012 with subsequent sedimentation (and erosion) processes especially during the flood period in June 2013. For a color version of this figure, see www.iste.co.uk/jacob/materials.zip*

Analyzing previous measures is an important basis for the planning and optimization of future measures. In the developed software tool, planning of dreading measures on individual shallow sections is possible as well. Figure 37.14 shows an example for planning future dredging measures for different LOSs with resulting water depth before and after dredging. Based on costs of previous measures and calculated dredging volume, it is also possible to estimate measure costs. However, as waterways are a serial system of shallow sections, a systematic analysis of all sections is necessary in order to calculate budgeting needs and measure priorities. Figure 37.15 shows an example of such a calculation for the 48 km free-flowing section of the river Danube east of Vienna including priorities, dredging volume, costs for individual sections and the entire river stretch. In addition, all previous dredging measures on individual sections are displayed, providing insight into measure impact duration. If these calculations are repeated for different combinations of fairway width and depth, it can be shown that empirical measure cost and transport cost surfaces correspond to the theoretical cost surfaces in Figure 37.1.

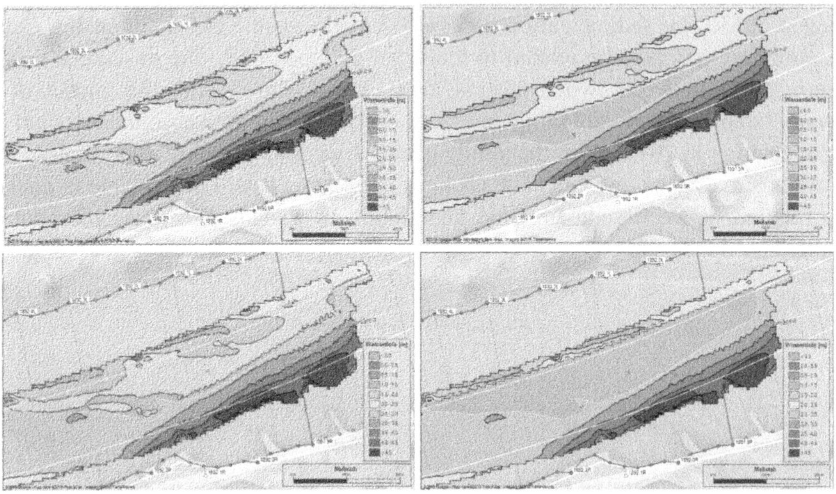

Figure 37.14. *Example for planning of dredging measures for different fairway width and a common fairway depth of 2.5 m with water depth based on LNWL before and after dredging at the shallow section Petronell–Witzelsdorf. For a color version of this figure, see www.iste.co.uk/jacob/materials.zip*

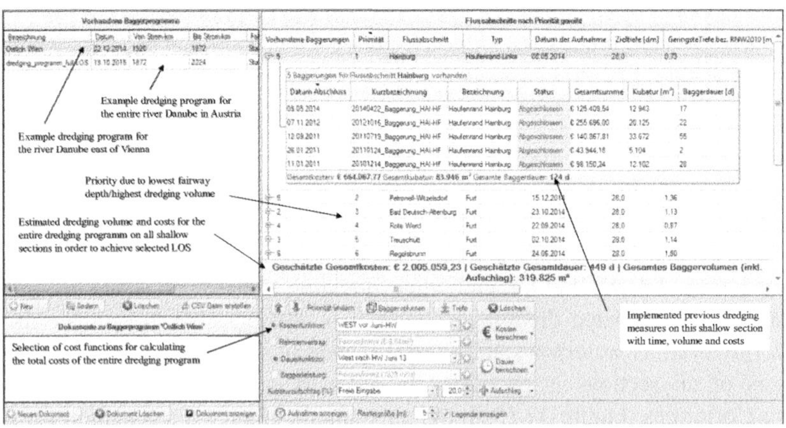

Figure 37.15. *Example for the calculation of a dredging program for a specific LOS-based fairway width and depth of 2.5 m (+0.3 m safety margin) with priorities, dredging volume and costs from 1,872–1,920 km of the River Danube*

37.5. Summary and outlook

Waterway asset management is a holistic multidisciplinary approach for the development, maintenance, rehabilitation and replacement of waterway assets based on a comprehensive lifecycle costing approach. With the availability at its core, the provided methodology allows for an optimization of measures for individual sections and entire transport routes alike. Furthermore, the presented approach is capable of including transport costs by allowing an optimization of fairway parameters for any given budget or minimizing both waterway agency and transport costs at once. Based on the analysis of the provided data, it was possible to verify the presented approach and combine all basic data in one unified database for a pilot stretch of 40 km. The automated analysis of riverbed characteristics, water levels and resulting availability is already feasible as well. The next steps of this pilot phase will include an extension of the database for all available information of the last 10 years on the entire Danube in Austria. Furthermore, an implementation of planning and assessment of operational and maintenance measures will be included as well. This first pilot phase was concluded in June 2015 with more challenging questions and promising results yet to come in a second phase of the project.

37.6. Bibliography

[ECC 12] ECCONET, Deliverable 3.3: Consequences of climate change for inland waterway transport, Report for the European Commission Directorate-General for Energy and Transport, Brussels, 2012.

[HOF 13] HOFFMANN M., Maintenance and rehabilitation of transport infrastructure as challenge for the future, Invited lecture, German at the planning days in Rust, 2013.

[PLA 07] PLANCO, Transport economics and environmental impact comparison of road, rail and waterway transport, Final Report, Magdeburg, Germany, 2007.

[SIM 04] SIMONER M., River engineering project eastof Vienna: environmental assessment, Final report navigation, Via Donau, Vienna, 2004.

[SCH 10] SCHWANZER J., HARTL S., Comparison of operating costs and primary energy use of inland navigation before and after river engineering projects on the river Danube, Internal Report, Via Donau, Vienna, 2010.

[VIA 13a] VIA DONAU, Danube navigation in Austria, Annual report, Vienna, 2013.

[VIA 13b] VIA DONAU, Manual on Danube Navigation, Vienna, 2013.

38

3D Numerical Simulation of Convoy-generated Waves and Sediment Transport in Restricted Waterways

Vessels such as towboats and barges can generate wake waves, cause draw-down and send waves back along riverbanks in narrow channel areas, and also contribute to sediment re-suspension and riverbank erosion.

In this chapter, a three-dimensional (3D) numerical model based on coupling a 3D Reynolds averaged Navier–Stokes (RANS) equation and a sediment transport model is presented to simulate ship-generated waves and re-suspended sediment transport as well as the erosive effects of propeller jets used by boats moving in waterways. Turbulence is taken into account by the k-ε model and the effects of propellers are discussed by considering the total thrust of the propulsion system. The results show that sediment re-suspension depends mainly on convoy speeds, the kinematics of convoy-generated wakes and the number of barges. Thus, the relationships between the re-suspended sediments and the advancing convoy speeds, the generated wakes, as well as the number of barges are studied.

38.1. Introduction

When a vessel moves in a narrow channel, the increased fluid velocity has a significant effect on the re-suspended sediment transport [ANT09]. Houser [HOU 11] found that the vessel-generated wakes (including drawdown and surge waves) have a much higher effect on re-suspended sediment than wind waves. John *et al.* [JOH 11] analyzed the relationship between the vessel's depression waves and the associated sediment re-suspension in Venice Lagoon (Italy). They found that the

---

Chapter written by Abdellatif OUAHSINE, Ji SHENGCHENG, Hassan SMAOUI, Philippe SERGENT and Nicolas HUYBRECHTS.

vessel-induced wakes led to suspended sediment concentrations above 400 mg/L, which is about 30 times higher than the average background concentration without ships. A one-dimensional (1D) numerical modeling of the sediment re-suspension induced by a moving vessel has been developed by Smaoui *et al.* [SMA 10] and validated by experimental results performed by Pham Van Bang *et al.* [PHA 09]. This numerical model shows a great agreement with the experimental results. It could handle the passage with two ships moving in series. The results of the 1D numerical model show a ls arge agreement with the experimental results; however, most 1D and 2D models do not allow us to simulate the sediment transport in the transverse direction of the ship's movement. Even if they are 3D models, they often do not take into account the hydro-sedimentary coupling associated with the passage of ships.

In this chapter, a 3D hydro-sedimentary model is introduced. The hydrodynamical model is based on the Reynolds averaged Navier–Stokes (RANS) equations (implemented in FLUENT 13.0) concerning the standard κ–ε model for turbulence processes. The sediment transport model (implemented as a user-defined subroutine) is based on the mass balance equations for the re-suspended sediment.

The chapter is organized as follows. Section 38.2 deals with the mathematical formulation of the problem, where the hydrodynamical model flow and the turbulence processes are introduced. In section 38.3, the 3D mass balance equation of the re-suspended sediment is presented where the associated boundary conditions for re-suspended sediment is given in section 38.4. In section 38.5, the computational procedures and results of the hydrodynamical model are presented. Section 38.6 is devoted to the numerical results and discussions concerning the sediment movement analysis, and section 38.7 gives conclusions drawn from the numerical investigation.

38.2. Numerical model and governing equations

The governing equations for mass and momentum conservation are RANS equations, which are given by

$$\frac{\partial \rho}{\partial t}+\frac{\partial(\rho u_i)}{x_i}=0 \quad i=1,2,3 \qquad [38.1]$$

$$\frac{\partial(\rho u_i)}{\partial t}+\frac{\partial(\rho u_i u_j)}{x_j}=\frac{\partial}{\partial x_j}\left[\mu\left(\frac{\partial u_i}{\partial x_j}+\frac{\partial u_j}{\partial x_i}-\frac{2}{3}\delta_{ij}\frac{\partial u_i}{\partial x_i}\right)\right]-\frac{\partial p}{\partial x_j}+\frac{\partial(-\rho \overline{u_i' u_j'})}{\partial x_j} \quad i=1,2,3 \qquad [38.2]$$

where $(x_i)_{i=1,2,3}$ are the Cartesian coordinates, t is the time, ρ is the density of the fluid, $(u_i)_{i=1,2,3}$ are the velocity components, p is the pressure, μ is the dynamic viscosity, δ_{ij} is the Kronecker delta, and $(-\rho\overline{u_i'u_j'})$ i,j=1,2,3 are the turbulent Reynolds stresses, which are modeled via the Boussinesq hypothesis given by:

$$-\rho\overline{u_i'u_j'} = \mu_t\left(\frac{\partial u_i}{\partial x_j}+\frac{\partial u_j}{\partial x_i}\right)-\frac{2}{3}\left(\rho k+\mu_t\frac{\partial u_i}{\partial x_i}\right)\delta_{ij} \quad i=1,2,3 \qquad [38.3]$$

where μ_t is the turbulent viscosityand k is the turbulent kinetic energy (TKE). We used the standard k–ε model for solving the problems. This turbulence model was introduced to improve the turbulence model based on the mixing length theory, as well as to find an alternative for algebraically prescribing turbulent length scales in moderate to high complexity flows. The standard k–ε model is given by:

$$\frac{\partial(\rho k)}{\partial t}+\frac{\partial(\rho u_i k)}{\partial x_i}=\frac{\partial}{\partial x_j}\left[\left(\mu+\frac{\mu_t}{\sigma_k}\right)\frac{\partial k}{\partial x_j}\right]+P_k-\rho\varepsilon+S_k \quad i=1,2,3 \qquad [38.4]$$

$$\frac{\partial(\rho\varepsilon)}{\partial t}+\frac{\partial(\rho u_i\varepsilon)}{\partial x_i}=\frac{\partial}{\partial x_j}\left[\left(\mu+\frac{\mu_t}{\sigma_\varepsilon}\right)\frac{\partial\varepsilon}{\partial x_j}\right]+C_{1\varepsilon}\frac{\varepsilon}{k}P_k-C_{2\varepsilon}\rho\frac{\varepsilon^2}{k}+S_\varepsilon \quad i=1,2,3 \qquad [38.5]$$

where ε is the dissipation of TKE, P_k is the production of TKE, S_k and S_ε are source terms and μ_t is the dynamic turbulent viscosity. P_k and μ_t are given as:

$$\mu_t=\rho C_\mu\frac{k^2}{\varepsilon} \qquad [38.6]$$

$$P_k=-\rho\overline{u_i'u_j'}\frac{\partial u_j}{\partial x_i} \qquad [38.7]$$

$C_{1\varepsilon}$, $C_{2\varepsilon}$, $C\mu$, σ_k and σ_ε are the constants of the standard k–ε model and given by: $C_{1\varepsilon}$= 1.44, $C_{2\varepsilon}$= 1.92, C_μ = 0.99, σ_k= 1.0 and σ_ε = 1.3.

38.2.1. *Free surface capturing*

The free surface between the air and the water phases is handled by using the volume of fluid method. Then, the interface tracking between the two phases is achieved by using the following equation:

$$\frac{\partial\alpha_i}{\partial t}+\boldsymbol{u}\cdot\nabla\alpha_i=0 \qquad [38.8]$$

where

$$\sum_{i=1}^{2} \alpha_i = 1 \quad \text{and} \quad \varphi = \sum_{i=1}^{2} \alpha_i \varphi_i \qquad [38.9]$$

where φ stands for the overall density and viscosity of the cell, i is the phase number, φ_i is the density and viscosity of phase i and α_i is its phase fraction. Equation [38.8] may be solved either by the classical up stream advection scheme or by the second-order Lax–Wendroff scheme for advection. In the presence of a strong gradient, it may be solved by using TVD schemes [OUA 99, SMA 08, ZOU 09].

38.3. Suspended sediment transport

The 3D equation of suspended sediment for each granulometric class is given by:

$$\frac{\partial \rho c}{\partial t} + \frac{\partial(\rho u_i c)}{\partial x_i} + \frac{\partial(\rho u_{si} c)}{\partial x_i} = \frac{\partial(-\rho \overline{u_i' c'})}{\partial x_i} \qquad [38.10]$$

where c is the mean SPM concentration, c' is the SPM fluctuation, u_i is the fluid velocity components corresponding to x_i, u_i' are the fluid velocity fluctuation components and u_s is the slip velocity components between sediment and fluid. Due to gravity, the velocity vector u_s is reduced to (0, 0, $-w_s$), where w_s is the particle's falling velocity:

$$w_s = \begin{cases} \sqrt{(s-1)gd_{50}/0.9}\left(A + \dfrac{4Bv}{d_{50}/0.9\sqrt{(s-1)gd_{50}/0.9}}\right) & if \quad 1\mu m < d_{50} < 1000\mu m \\ 1.1\sqrt{(s-1)gd_{50}} & if \quad d_{50} > 1000\mu m \end{cases} \qquad [38.11]$$

with $s = \rho_s/\rho$, where ρ_s is the sediment density, ρ is the fluid density, d_{50} is the median diameter of the particles, constants A and B are, respectively, equal to 0.954 and 5.12 for quartz sand [JIM 03], v is the fluid kinematic viscosity and g is the acceleration due to gravity. The turbulent sediment flux (correlations between fluctuations of turbulent velocity and fluctuations of the SPM) is modeled as follows:

$$-\overline{u_i' c'} = \varepsilon_{si} \frac{\partial c}{\partial x_i} \qquad [38.12]$$

The sediment mixing coefficient ε_{si} is assumed to be proportional to the eddy kinematic viscosity v_t. Then:

$$\varepsilon_{si} = \frac{v_t}{\sigma_c} \qquad [38.13]$$

where $\sigma_c = 0.5$ is the turbulent Schmidt number.

38.4. Boundary conditions

For the hydrodynamic model, we assume that the x-component of the fluid velocity u is equal to the convoy velocity V, where $V = 1.151$ m/s at inlet boundaries. Herein, the free surface is assumed to be horizontal and normal to the direction of gravity g. The water volume fraction α is equal to 0.5 at the free surface. Below the free surface level α is set to 1.0, while α is 0.0 above the free surface level. The symmetry condition is used in the symmetry plane as shown in Figure 38.2, which means that the normal velocity component and the normal gradients of all flow variables are null at the symmetry plane. Furthermore, the shear stress is equal to zero at the symmetry boundary. All the other boundaries are set to the non-slip wall conditions [JI 12].

For the transport model, the distribution of the SPM concentration in the suspended transport is controlled by the convection diffusion equation [38.10], while the bed load transport rate can be solved by a transport formula. Elsewhere, a near-bed equilibrium SPM concentration c_b is specified by the following equation [VAN 87]:

$$c_b = \beta \frac{d_{50}}{\delta} \frac{T^{1.5}}{D^{0.3}} \qquad [38.14]$$

where d_{50} is the median grain diameter (=70 μm). D is the dimensionless particle parameter:

$$D = \left(\frac{(s-1)\,g}{v^2} \right)^{1/3} d_{50} \qquad [38.15]$$

where β is the calibration coefficient, δ is the reference level, T is the non-dimensional excess of the bed shear stress:

$$T = \frac{\tau_b - \tau_b^{cr}}{\tau_b^{cr}} \qquad [38.16]$$

where τ_b is the instantaneous bed shear stress and τ_b^{cr} is the critical bed shear stress [SMA 12]. In order to evaluate the threshold for the critical shear stress on the bottom, we used the following relationship [PAP 09]:

$$\theta_{cr} = \frac{0.186}{1+0.2\varpi} + 0.0445(1-0.988\,e^{-0.01\varpi}) \quad (0.1 < Re_* < 10^6) \qquad [38.17]$$

where Re_* is the grain Reynolds number, θ_{cr} is the Shields entrainment function and ϖ is Yalin's approach [PAP 09]. They are defined as follows:

$$Re_* = \frac{u_* d_{50}}{\nu}, \quad u_* = \sqrt{\frac{\tau_b}{\rho}} \qquad [3.18]$$

$$\theta_{cr} = \frac{\tau_b^{cr}}{(\rho_s - \rho)\,g d_{50}}, \quad \varpi = \left[\frac{(s-1)\,g d_{50}^3}{\nu^2}\right]^{1/2} \qquad [38.19]$$

38.5. Computational procedures and hydrodynamical model results

The set of equations [38.1]–[38.9] is solved by the finite volume method with arbitrary hexahedral shapes. Since the values of the variables are stored in the center of each cell, a procedure of interpolation, based on a second-order scheme, is then used to assign values to the faces of the control volume. The SIMPLE algorithm (Semi-Implicit Method for Pressure-Linked Equations) was used for pressure velocity coupling. For the added scalar quantity SPM, the first-order implicit scheme is selected for time discretization and the QUICK scheme for convective and diffusive terms.

The computational domain was built using an unstructured mesh in order to provide a sufficiently high resolution in the vicinity of the moving vessel trajectory. On the contrary, a coarse mesh resolution is used away from the areas of interest, where the detailed spatial discretization is not required [JI 12, JI 14].

Top width (B)	Bottom width (b)	Bank slope (α)	Calm wetted section (A_c)	Pusher length (L_p)	Barges length (L_b)	Barges number (N)	Convoy length (L)
5.4 m	3.6 m	tan α=1/2	2.025 m^2	3.00 m	7.65 m	1	18.3 m
Water depth (h)	Wetted perimeter (X_c)	Hydraulic radius		Convoy width (w)	Draught (T)	Cross-section (A_c)	
0.45 m	5.61 m	0.361 m		1.14 m	0.1 m	0.114 m^2	

Table 38.1. *Geometrical parameters of the channel and the convoy [COM 97b]*

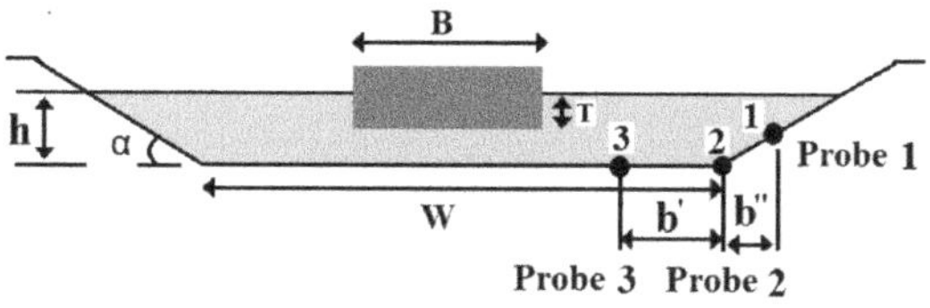

Figure 38.1. *Diagram of the convoy and location of the probes used for the comparison*

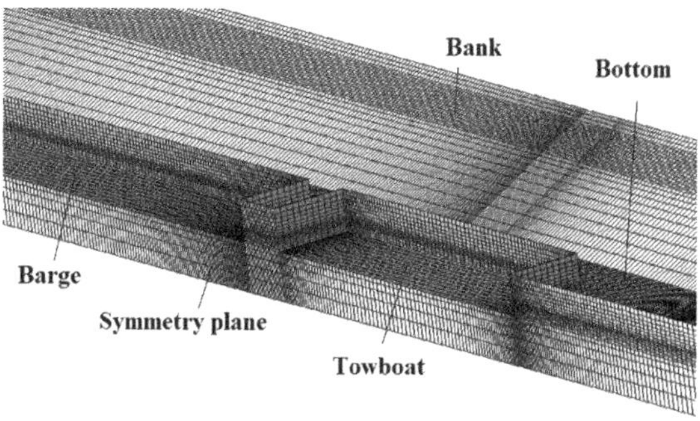

Figure 38.2. *Meshes on hull surfaces and on the boundaries*

The schematic representation of the model geometry for the hydrodynamic model is shown in Figure 38.1 and the hull surface and boundary meshes are shown in Figure 38.2. Figure 38.3 shows the computed wave pattern generated by the barge and towboat, and Figure 38.4 shows the comparison between the computed and experimental results [COM 97b] of the ship wave profile at probe 2 (along the line at y = 1.8 m). It shows that the computed drawdown H_d^{comp} = 0.025 m is 10.7% lower than the experimental drawdown H_d^{expt} = 0.028 m. For the secondary waves, the computed maximum wave height H_m^{comp} =0.022 m is 18.5% higher than the measured H_m^{expt} = 0.027 m [COM 97b].

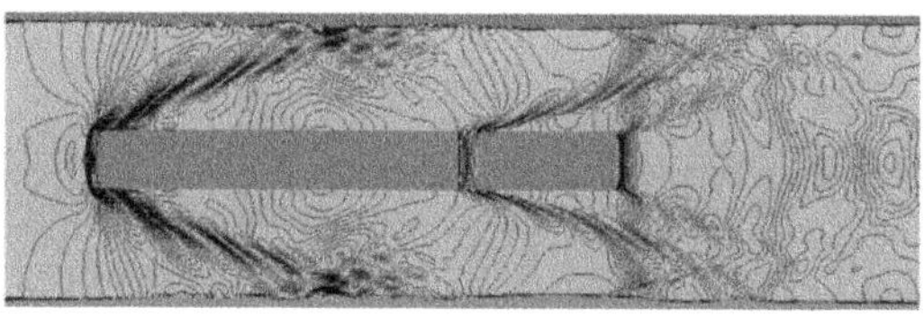

Figure 38.3. *Simulated free surface including a barge and a towboat*

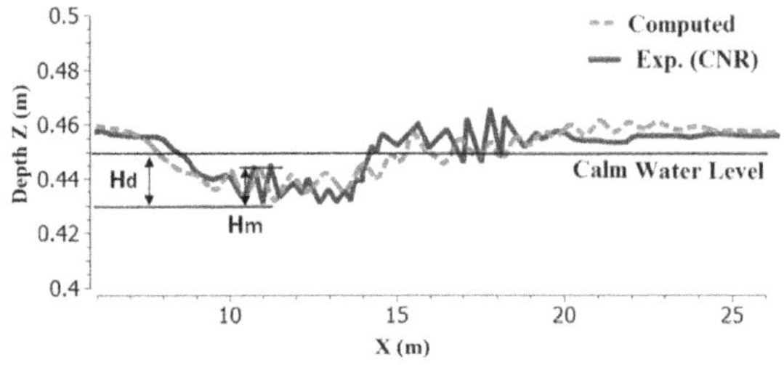

Figure 38.4. *Wave profile at probe 2 (see Figure 38.1): comparison with experimental data*

38.6. Navigation influences on sediment transport

In order to ensure the accuracy of the hydrosedimentary model, the used grid density is bigger than the validated hydrodynamic model (Table 38.2). We also used the same boundary conditions as for the hydrodynamic model, given in section 38.4.

Ship model	Length L_b	Beam B	Depth D	Taught T	Ship speed V_b (m/s)
	3m	0.52m	0.39 m	0.2 m	0.5–1.3
Waterway scales	Length L	Width W	Depth h	Water velocity	Froude number Fr
	30 m	2.0–5.0 m	0.4 m	0.1 m/s	0.20–0.66

Table 38.2. *Convoy model and waterway related to sediment transport*

The origin of the coordinates is set at the center of the ship for the calm free-surface level. The x, y and z axes are directed in the opposite direction of ship motion, transversely and upward, respectively. The impact of the surface wakes on the sediments transport is examined by analyzing the evolution of the SPM re-suspension at a fixed point $z/h = -0.975$, which is located closely to the bottom. This SPM evolution is supposed to depend on the ship speeds and the wave height.

Figure 38.5 gives the relationship between the SPM_{max} and Froude number. We found that the maximum SPM is proportional to the Froude number (or the advancing ship velocities in our case). Figure 38.6 shows the SPM distribution along lines L_1, L_2 and L_3 on the symmetry plane for $V_b = -0.6$ m/s. Three peaks of SPM have been generated by the moving convoy. The first peak is produced by the head of the first barge, whereas the second and third peaks are induced by the gap between the two barges and the stern of the second barge, respectively.

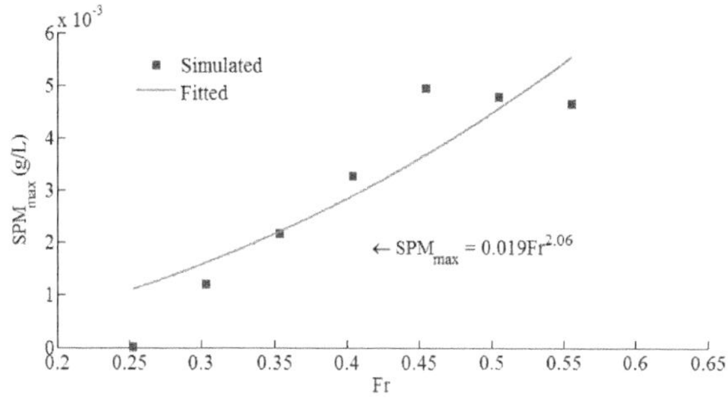

Figure 38.5. *Maximum SPM versus Froude number*

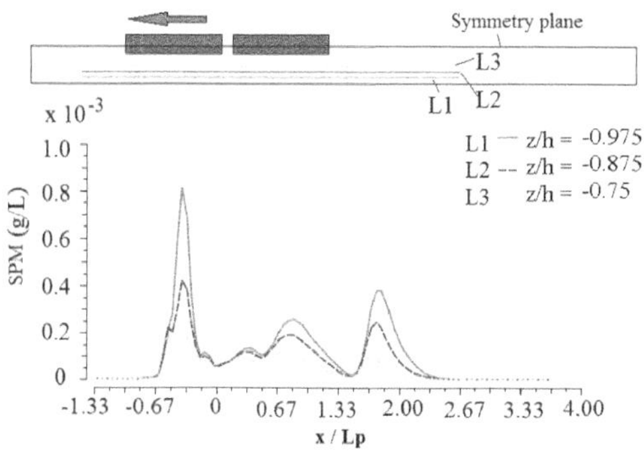

Figure 38.6. *SPM distribution along lines L_1, L_2 and L_3 on the symmetry. For a color version of this figure, see www.iste.co.uk/jacob/materials.zip*

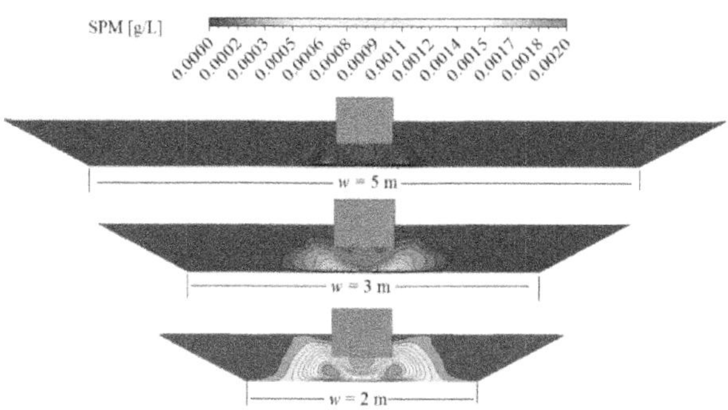

Figure 38.7. *SPM distribution on trans-section for different blockage coefficient (C_b) for V_b = –0.8 m/s. For a color version of this figure, see www.iste.co.uk/jacob/materials.zip*

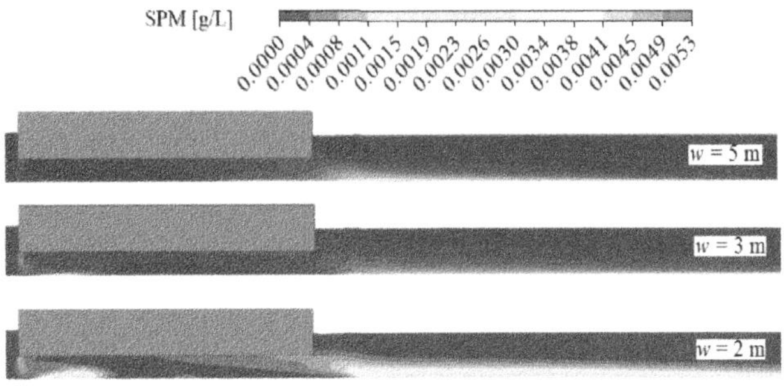

Figure 38.8. *SPM distribution on the symmetry plane for different value of blockage coefficient (C_b) for V_b = –0.8 m/s. For a color version of this figure, see www.iste.co.uk/jacob/materials.zip*

Figures 38.7 and 38.8 show the SPM distribution on the trans-sections and symmetry planes, respectively, for different values of the blockage coefficient $\boldsymbol{C_b}$ corresponding to canal bottom width $\boldsymbol{w}$ = 5 m, 3 m and 2 m. The blockage coefficient $\boldsymbol{C_b} = \boldsymbol{A_{ship}}/\boldsymbol{A_{canal}}$, where $\boldsymbol{A_{ship}}$ and $\boldsymbol{A_{canal}}$ denote the cross-section area below the free surface of the ship and the canal, respectively. These figures show that the SPM become more diffusive with increasing $\boldsymbol{C_b}$.

Figure 38.9 shows the time history of the SPM value at the same point z/h = –0.975 on the symmetry plane for the different values of the blockage coefficient C_b. From Figures 38.7–38.9, we can see that the blocking coefficient has a significant influence on the increase and distribution of SPM in the confined channel.

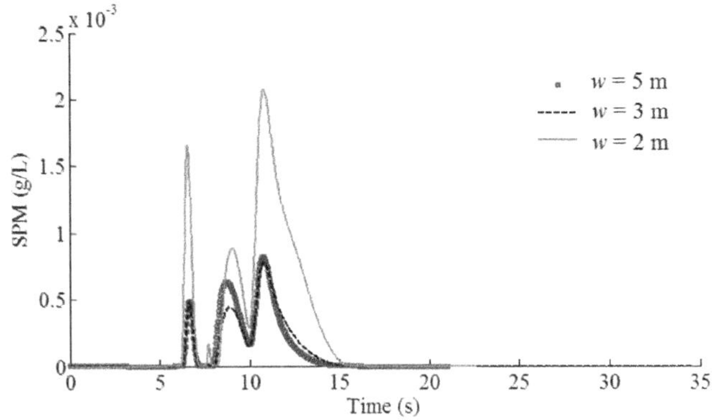

Figure 38.9. *SPM time history at a point fixed z/h = –0.975 on the plane symmetry for V_b = –0.8 m/s*

38.7. Conclusions

3D hydrosedimentary is presented to identify the relationship between the sediment movement and the pattern of ship-generated waves around and far away from the ship. The numerical results highlighted the relationships between the maximum SPM and the speed of ship. It shows that the areas of sediment concentration and transport depend mainly on the position and the speed of the ship in the waterways, kinematics of ship-generated waves and return flows. For the case with two barges, two additional peaks were generated with reference to the case with only one barge.

38.8. Bibliography

[ANT 09] ANTS E., VICTOR A., MADIS L., "Monitoring wave-induced sediment resuspension", *Estonian Journal of Engineering*, vol. 15, no. 3, pp. 196–211, 2009.

[COM 97a] COMPAGNIE NATIONALDU RHONE (CNR), Laboratoire d'Hydraulique et d'Essais de Matériaux. C753-DLLAB 97-490, 1997.

[COM 97b] COMPAGNIE NATIONALDU RHONE (CNR), Etude Des Ondes de Batillage sur Modèle Physique au 1/30 Phase 1: Exploitation des données relatives aux essais réalisés entre juillet et décembre 96, 1997.

[JOH 11] JOHN R., LUC Z., KLAUS R. *et al.*, "Characteristics of ship's depression waves and associated sediment resuspension in Venice Lagoon, Italy", *Journal of Marine Sysyems*, vol. 85, pp. 45–56, 2011.

[JIM 03] JIMÉNEZ J.O., MADSEN O.S., "A simple formula to estimate settling velocity of natural sediments", *Journal of Waterway, Port, Coastal and Ocean Engineering*, vol. 129, no. 2, pp. 70–78, 2003.

[HOU 11] HOUSER C., "Sediment resuspension by vessel-generatedwaves along the Savannah River, Georgia", *Journal of Waterway, Port, Coastal, and Ocean Engineering*, vol. 137, no. 5, pp. 246–257, 2011.

[OUA 99] OUAHSINE A., SMAOUI H., "Flux-limiter schemes for oceanic tracers: application to the English Channel tidal model", *Journal of Computer Methods in Applied Mechanics and Engineering*, vol. 179, pp. 307–325, 1999.

[OUA 12] OUAHSINE A., SMAOUI H., MEFTAH K. *et al.*, "Numerical study of coastal sandbar migration, by hydro-morphodynamical coupling", *Journal of Environmental Fluid Mechanics*, vol. 13, no. 2, pp. 169–187, 2012.

[JI 12] JI S-C, OUAHSINE A., SMAOUI H. *et al.*, "3D numerical simulation of convoy generated waves in a restricted waterway", *Journal of Hydrodynamics*, vol. 24, no. 3, pp. 420–429, 2012.

[JI 14] SHENGCHENG J., OUAHSINE A., SMAOUI H. *et al.*, "3D numerical modeling of sediment resuspension induced by the compounding effects of ship-generated waves and the ship propeller", *Journal of Engineering Mechanics*, vol. 140, no. 6, pp. 04014034, 2014.

[PHA 09] PHAM VAN BANG D., SERGENT P., HISSEL F. *et al.*, "Vagues et érosion des sédiments générés par le passage des bateaux de navigation intérieure: mesures insitu et analyses", *Revue Technique Maritime et Fluviale*, vol. 2, pp. 18–24, 2009.

[PAP 01] PAPHITIS D., "Sediment movement under unidirectional flows: an assessment of empirical threshold curves", *Coastal Engineering*, vol. 43, pp. 227–245, 2001.

[SMA 08] SMAOUI H., ZOUHRI L., OUAHSINE A., "Flux limiting techniques for simulation of pollutant transport in porous media: Application to ground water management", *Journal Mathematical and Computer Modelling*, vol. 47, pp. 47–59, 2008.

[SMA 10] SMAOUI H., OUAHSINE A., PHAM VAN BANG D. *et al.*, "Numerical modeling of the sediment re-suspension induced by boat traffic", in GINSBERG S.S. (ed.), *Sediment Transport*, In Tech Publisher, Croatia, 2010.

[SMA 12] SMAOUI H., OUAHSINE A., "Extension of the skin shear stress Li's relationship to the flat bed", *Journal of Environmental Fluid Mechanics*, vol. 12, no. 3, pp. 201–207, 2012.

[VAN 87] VAN RIJN L.C., Mathematical modeling of morphological processes in the case of suspended sediment transport, Doctorate thesis, Water loop kunding Laboratorium delft, 1987.

[VAN 90] VAN RIJN L.C., "Field verification of 2-D and 3-D suspended-sediment model", *Journal of Hydraulic Engineering*, vol. 116, no. 10, pp. 1270–1288, 1990.

[ZOU 09] ZOUHRI L., SMAOUI H., CARLIER E. *et al.*, "Modelling of hydrodispersive processes in the fissured media by flux limiters schemes (chalk aquifer)", *Journal Mathematical and Computer Modelling*, vol. 50, nos. 3–4, pp. 516–525, 2009.

PART 6

Climate Resilient Roads

39

Potential Impact of Climate Change on Porous Asphalt with a Focus on Winter Damage

This chapter investigates the impact and adaptation options of climate change on porous asphalt (PA) roads, specifically for the case of winter weather (freeze–thaw cycles) and road damage in the Netherlands. Changes in weather patterns pose a threat to the serviceability and long-term performance of roads, as up to half of road maintenance costs are attributable to weather stresses. PA is of particular concern in the Netherlands, where its use has become mandatory, primarily for environmental (noise reduction) concerns. In recent winters, ravelling and pothole damage have increased the discussion about cold weather performance of PA and the potential challenges of changing winter weather patterns. Current climate change impact research often produces results on a systemic, macroscale, and less is known about the regional impact to specific road types. To address this, we examine the correlation between historic winter weather and PA pavement performance, which is particularly sensitive to the freezing/thawing phenomena. That relationship is combined with Dutch regional climate models and used to analyze the potential physical and economic impacts of adapting to future climate change. This has implications on maintenance, design, and long-term planning of the road network in the Netherlands.

39.1. Introduction

This chapter investigates the impact and adaptation options of climate change on porous asphalt (PA) roads, specifically for the case of winter weather (freeze–thaw cycles) and road damage in the Netherlands. Changes in weather patterns pose a threat to the serviceability and long-term performance of roads, as up to half of road maintenance costs are attributable to weather stresses [NEM 12]. PA is of particular

Chapter written by Kyle KWIATKOWSKI, Irina STIPANOVIC OSLAKOVIC, Andreas HARTMANN and Han TER MAAT.

concern in the Netherlands, where its use has become mandatory, primarily for environmental (noise reduction) concerns. In recent winters, raveling and pothole damage have increased the discussion about cold weather performance of PA. Current climate change impact research often produces results on a systemic, macro-scale, and less is known about the regional impact to specific road types. To address this, we examine the correlation between historic winter weather and PA pavement performance, which is particularly sensitive to the freezing/thawing phenomena. That relationship is combined with Dutch regional climate models (RCMs) and used to analyze the potential physical and economic impacts of adapting to future climate change, which has implications on maintenance, design and long-term planning of the road network in the Netherlands.

The research is conducted using a combination of empirical data analysis, climate and cost modeling. Historic winter maintenance data are recorded by Rijkswaterstaat, the ministerial road authority of the Netherlands, and winter weather data are provided by KNMI, the Dutch meteorological institute. By analyzing these databases, for correlations between weather and damage, a stressor–response function is developed that relates the frequency of freeze-thaw cycles to the frequency of pavement damage. The stressor–response function is incorporated into the previously developed Infrastructure Planning Support System (IPSS), which analyzes the impacts and costs of different adaptation options for climate change and infrastructure [CHI 11]. To increase the granularity of analysis and produce regional (provincial) level results, downscaled Dutch RCMs are used. The results of IPSS are presented as an annual economic impact (costs) of each adaptation strategy from the current year until 2100.

The research is currently in progress. The changes in winter climate and impact on road damage are expected to vary regionally within the Netherlands. The IPSS results will highlight regions where proactive adaptation may create costs savings, as well as regions where adaptation is not financially advantageous. Discussion will contain practical implications for Rijkswaterstaat, both for short- and long-term planning and policy decision making. Of particular interest are the long-term viability and maintenance considerations of PA pavement in the future climate conditions. Adaptation strategies, pavement characteristics and cost data are based on actual maintenance and design practices of the organization, which will result in output and recommendations that are both relevant and actionable for the Netherlands Ministry of Transportation.

39.1.1. *Climate change and road national authorities*

Climate change adaptation research has been a fast growing field as scientists and practitioners now acknowledge that even with mitigation the planet will experience certain unavoidable levels of climate change [IPC07]. While the questions of how much and when the changes will occur are still being debated, there is a search for improved understanding of the potential impacts on transport infrastructure from future climate change. There are also several detailed research works that have analyzed specific material responses to climate change, such as pavement implications [MIL 09]. As a larger quantity of data is produced, it is important to examine the scientific robustness, but also the relevance and usability of that information for the infrastructure designers and operators.

Organizations, often public organizations, are now and will continue to be responsible for planning and implementing climate change adaptation measures in the future. This is especially true for the transportation infrastructure sector, which is primarily managed by public agencies and requires long-term planning for both design and maintenance. In order to anticipate and plan for changes that may be required in the uncertain future, these agencies require information that is more detailed than sector-wide analysis but less in-depth than material property studies. The research project described in this chapter is based on those efforts by combining economic modeling and material science while also maintaining an organizational perspective. The work intends to advance the science of climate change adaptation research through improvements to the cost modeling methodology, while also producing results that are more relevant and implementable at the organizational level.

Within the transportation sector highway, organizations face a difficult challenge, as roads are a particularly vulnerable part of infrastructure. For example, as much as half of road maintenance costs are attributable to weather stresses [NEM 12]. Roads are designed to operate with minor variability in weather but long-term changes in climate are not accounted for in current planning and design standards. Changes in temperature and precipitation may result in positive outcomes, in the form of warmer winters for example, but also potentially severe negative outcomes from increased rainfall and higher temperatures [PET 08].

It is important for transportation (highway) agencies to understand and adapt to these impacts for the end-users safety, for their own organizational benefit (e.g. cost and resource efficiency) and also because of the role that roads play in national commerce and international trade. Consequently, as the need for evaluation of adaptation options and costs has been established, research on climate change and roads has become increasingly quantitative, highlighted in the United States by Chinowsky *et al.* [CHI 13b]. However, these economic modeling results are not intended to be organization specific but rather to inform at the national policy level [JOT 10].

39.1.2. *Climate change adaptation and road infrastructure*

Climate change research has evolved from focusing primarily on mitigation to include the exploration and evaluation of adaptation options. While there continues to be debate over the merits and means of mitigation, there is general consensus that some form of adaptation will be required to address the amount of climate change that is now unavoidable [IPC 07]. Economic modeling of the impact and adaptation costs has been completed in many industry sectors, especially in agriculture and water resources [TOL 02]. This broad quantification of the potential impacts is an important tool for economists and policymakers. But as Jotzo states in his report on the limitations of economic modeling for climate change adaptation, more detailed sector-specific modeling is required to enable local adaptation action [JOT 10].

This is true within the transportation sector and there are an increasing number of studies [JAR 10, PET 08, TRB 08] that examine the risks of climate change to transportation infrastructure. Specifically, higher temperatures and increased precipitation will result in the acceleration of road degradation, primarily through rutting and ravelling. The quantification and modeling of cost impacts is being researched [CHI 13b, NEM 12]; however, there are still few detailed studies, such as Mote *et al.* [MOT 12] in the United States, which analyze climate and road characteristics that are unique to a specific organization. As stated in a report on climate change impact to European Union rails and roads, "Both vulnerability and adaptation costs would need to be assessed under a much higher spatial resolution" [NEM 12].

Sector-specific modeling has been completed for road transportation using the IPSS. This support system was developed by the Institute for Climate and Civil Systems at the University of Colorado Boulder to model the cost impact of adaptation options for climate change and road infrastructure. In addition to the social benefits of roads in developing countries, this work also detailed the importance of roads to the overall economic success of a country [CHI 11]. This is especially true for the highway system of the Netherlands.

39.2. Porous asphalt

39.2.1. *PA background*

PA pavement is characterized by a high percentage of interconnected voids in the top friction layer of the pavement. This creates high permeability and is also capable of reducing tire noise. The maximum aggregate size used in single layer porous pavements is 16 mm with a layer thickness of 50 mm and a built-in air void of 20%. Standard bitumen without modification is used for single layer porous pavements. In 2007, the bitumen percentage was increased from 4.2 to 5.2% and this is considered to increase the lifetime of the pavements by around 20%. For two-layer PA, the maximum aggregate size in the top layer is normally 8 mm [VEJ 12].

A major motivation for its use in the Netherlands is that the cost and space efficiency is higher than sound barriers [ALV 06]. PA pavements also have high resistance to rutting [MIR 09]. Despite these benefits of PA, there are drawbacks that are important to consider in the context of climate change. PA has a limited lifespan that is shorter than most alternative pavement types [MIR 09], it is susceptible to increased ravelling and accelerated aging from rainwater and de-icing salt [SU 13], and it has been shown that noise reduction effectiveness can be significantly reduced with age and clogging of pores [BEN 05]. These will be important for Road Weather Station (RWS) to consider when evaluating the long-term effectiveness of PA pavement in a changing future climate.

As described previously, the Dutch include environmental concerns in their road network planning. This includes the impact of road construction to the surrounding environment, as well as quality of life for nearby citizens. With a small land area and dense road network, residents in the Netherlands often live near highways and noise reduction is therefore an environmental consideration for road and land use planning. As the population grows and the use of automobiles increases, RWS has made noise reduction a priority [VAN 02, HUU 10]. As a result, as well as for additional reasons such as drainage, the Dutch road network is now constructed with approximately 90% PA pavement. This reliance on PA pavement increases the importance of climate change considerations in long-term planning of the Netherlands road network.

39.2.2. *Historical behavior of PA in winter*

The expected lifetime of single layer PA is 1–2 years less than for dense asphalt concrete. In the Netherlands, experience has shown that the approximate service life for PA wearing courses is 10 years versus 12 years for dense asphalt concrete [VAN 90]. The most often cited mode of deterioration for PA is ravelling, which primarily occurs during the winter season. PIARC (1993) states that rutting or cracking rarely appears in PAs, except for reflection cracks, and that loss of material is one of the major reasons for road maintenance.

October 2009 and November 2010 winters were severe, characterized by heavy snowfall and frost as well as many freeze-thaw cycles. This resulted in damage to Dutch roads, which was observed on all types of pavements. But as 90% of the high-way network has PA, the damage was mainly experienced by these pavement types.

The damage observed consisted mostly of ravelling, potholes and material loss at longitudinal joints between PA (see Figure 39.1).

Additionally, damage was mainly on pavements at the end of their expected lifetime. Old PA is more sensitive to frost damage in hard winters because the bitumen hardens as the pavements gets older. As noted previously, older pavement is also shown to be less effective at noise reduction. Small cracks develop on pavements with aged and hardened bitumen. Water penetrates into these cracks and in periods of frost, the water freezes and aggregate loosens from the pavement surface.

Figure 39.1. *Frost damage on porous pavement [RWS 12]*

Potholes also contribute to accelerated road degradation in PA. For instance, emergency lanes become clogged and when they are not cleaned, water collects in the pores of the pavement in the summer period. This can result in the mortar/bitumen stripping from the aggregate and the damage can be accelerated by freeze-thaw cycles in the winter. Most of the frost damage has occurred on old pavement sections, which according to the Dutch pavement management system are already scheduled for replacement. A delay in planned maintenance work, especially on PA, can cause additional frost problems [VEJ 12].

A severe winter is statistically seen only once every 10 years in the Netherlands. Therefore, frost damage is considered by the national road administration as a calculated risk that happens, on average, once in 10 years. As seasonal climate patterns change, this risk will also change. Rijkswaterstaat recognizes the need for more fundamental research into the causes of frost damage and the results may make it possible to develop better frost resistant mixtures of PA. In 2010, Rijkswaterstaat published guidelines for handling asphalt pavements including porous pavements [RWS 10]. One method for improving frost resistance is through the use of more bitumen in pavement mixes. This will increase the cost of PA, however, and may not be effective under different climate conditions in the future.

39.2.3. *Temperature monitoring*

The Road Weather Information System (RWIS) is a continuous monitoring system for early warning of slipperiness on highway pavement surfaces. The RWIS system consists of 285 (RWSs combined with small local meteorological measuring stations. These stations are used to produce warnings of potentially upcoming slippery situations based on road surface temperature, resistance (dry, wet, or salt), dew point and precipitation.

For this project, we have analyzed road surface temperatures at two locations on highway A10, with embedded temperature sensors, as shown in Figure 39.2. Road temperatures are measured in the surface layer and air temperature is measured beside the road in the small weather station. In order to establish correlations between official meteorological stations, road temperatures were initially correlated to the air temperature immediately adjacent to the road. This was then correlated to temperatures measured at the nearest meteorological weather station, in this case KNMI Schiphol weather station. Temperature measurements were correlated for the period of 3 years. Results are presented in Figures 3.3(a)–(d).

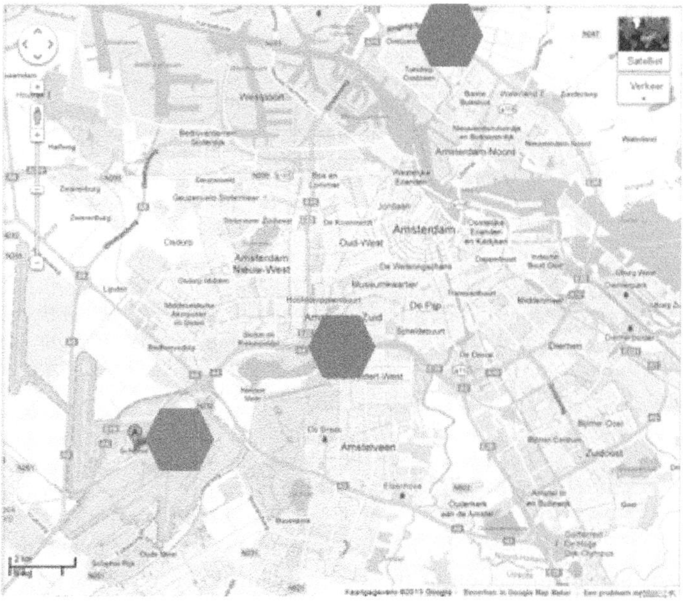

Figure 39.2. *Location of road sensors on highway A10 and KNMI weather station*

The goal of this part of the research was to establish numerical correlations that can be used for translating air temperatures into pavement temperatures, where no direct measurement of the road surface temperatures exist. The established correlations will be then used in predicting frost damage that is dependent on road temperature. As seen in Figure 39.3, the temperature conversion equations have very high correlation factors and are appropriate for the modeling purposes.

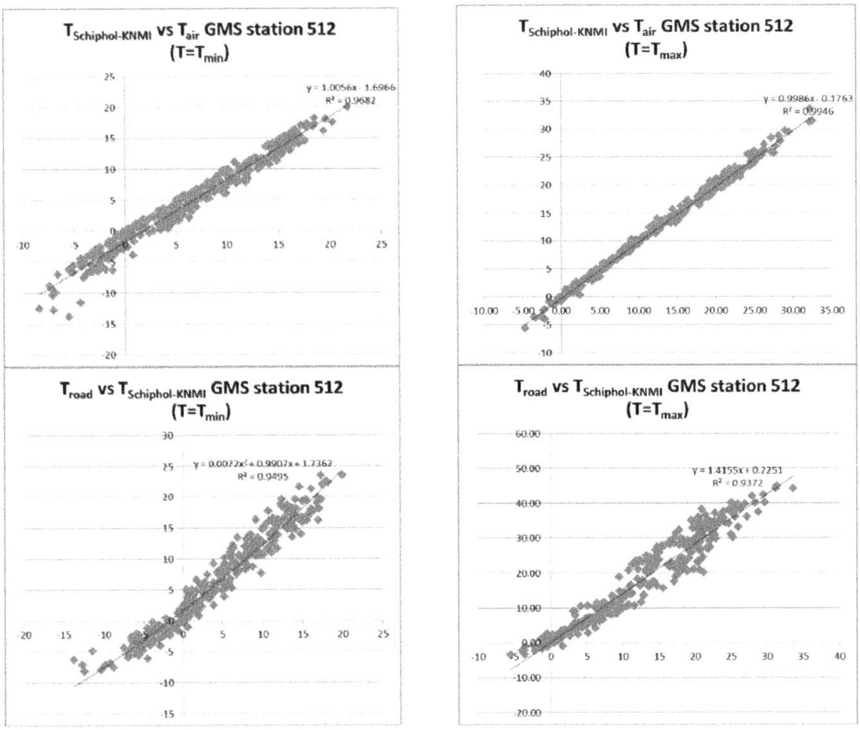

Figure 39.3. *Results of temperature analysis: a) correlation between minimum air temperature measured beside the road and with official KNMI station; b) correlation between maximum air temperature measured beside the road and with official KNMI station; c) correlation between minimum road temperature and air temperature measured at the KNMI station and d) correlation between maximum road temperature and air temperature measured at the KNMI station*

39.3. Modeling of climate change impact

The research is conducted through a combination of climate modeling, empirical data analysis and modification of the existing climate change adaptation modeling system, IPSS. The model has been used for several case studies, including countries in Europe [CHI 11]. Until now, the model has been used primarily for developing countries and focused on temperature and precipitation stressors. Using the Netherlands as a case study, the IPSS model is expanded to include country-specific road characteristics (see Table 39.1), freeze-thaw impacts, and increased granularity of analysis. Incorporating country-specific road design and maintenance data strengthen the relationships between climate stressors (freeze-thaw cycles) and physical road response (damage).

Road Type	Characterisitc	**RL**	**LL**
DAB	Service life	12 years	18 years
	Construction costs (compared to DAB)	1	
	Maintenance costs (compared to DAB)	1	
ZOAB	Service life	11 years	17 years
	Construction costs (compared to DAB)	1.02	
	Maintenance costs (compared to DAB)	1.05	
ZOAB TW	Service life	9 years	13 years
	Construction costs (compared to DAB)	1.12	
	Maintenance costs (compared to DAB)	1.80	

Table 39.1. *Porous asphalt road characteristics by type [VAN 05]*

39.3.1. *Infrastructure Planning Support System*

Cost modeling is completed using IPSS, the climate adaptation modeling system designed by the Institute for Climate and Civil Systems (iCliCS) at the University of Colorado Boulder. Previous versions of the IPSS model analyzed paved, gravel, and dirt roads primarily for developing countries. This research expands on the paved road portion of the model by including freeze-thaw analysis and focusing on PA road type.

Climate data are incorporated into the analysis on a 0.5° by 0.5° (longitude/latitude) grid for temperature and precipitation changes through 2100. Using higher resolution RCMs, additional analysis is performed at the 0.25° by 0.25° scale. The forecasted climate change within each grid cell is applied to the total length of roads located in that same grid. This process quantifies the physical impact on road performance and design life into costs of adaptation, as outlined below.

The model uses thresholds and stressor–response functions to predict the impact that changes in climate stressors, in this case the total number and frequency of freeze-thaw cycles, will have on a road. The functions compare changes in quantities of freeze-thaw cycles to historic road and weather data, which are used to set thresholds at which freeze-thaw is shown to begin negatively affecting road performance. If the future freeze-thaw patterns exceed these thresholds, changes in design or maintenance will be required to "climate-proof" the road against the change in climate. In [CHI 12], more detailed explanation of the functions, stressor–response methodology and thresholds can be found. Chinowsky *et al.* [CHI 13a] also provide background on the impact functions and underlying assumptions of the model. For this case study, the basic assumptions and process of the model remain the same. However, Dutch road design parameters – including cost, lifespan and pavement type – replace the original IPSS road inputs. Combined with the downscaled climate modeling, this modifies the system to produce results that are organization specific and on a finer spatial scale than previously attained.

The IPSS results are calculated based on two alternate adaptation strategies. A "no adapt" strategy calculates costs based on increased maintenance that is required to preserve the intended design life of the road, without altering the initial design of that road. The "adapt" strategy calculates costs for altering the initial design of a road with increased climate resiliency during its lifespan, rather than preserving the initial design life exclusively through maintenance. The adaptation evaluation is performed at the beginning of each road's design life. If the climate model forecasts that a threshold will be crossed (i.e. the number of freeze–thaw cycles has increased) during a road's design life, the cost for adapting and not adapting isthen calculated. For this project, it was crucial to use data from RWS in the Netherlands in order to produce quantitative results that are based on actual costs, policy, and procedure, making them more actionable for the infrastructure planners.

39.3.2. *Climate modeling*

In addition to the global climate models (GCMs) that IPSS typically utilizes, this case study includes downscaled climate data that arespecific for the Netherlands' region of Europe.

39.3.2.1. *Climate description of the Netherlands*

The Netherlands has a temperate maritime climate, with cool summers and moderate winters. The country is small and, as a result, there is little variation inland; although the influence of the sea is noticeable in the western part of the country. Daytime temperatures vary from 2 to 6°C in wintertime and 17 to 20°C in summertime. Precipitation is distributed equally throughout the year.

39.3.2.2. *RCM simulations*

RCMs are a complementary research method to the coarser resolution GCMs. High resolution is one key advantage of RCMs (spatial resolution of 25–50 km) compared with GCMs (spatial resolution at best around 100–200 km), especially in regions with variable land forms or characteristics. The quality of a RCM simulation, with a spatial resolution of 25–50 km, is dependent onthe RCM itself and onthe driving GCM.

The ENSEMBLES project was a large research program founded by the European Commission in 2004. The main aim, and core, of the ENSEMBLES project was running multiple climate models ("ensembles") with the aim to produce a range of future predictions assessed to decide which of the outcomes are more likely (probable) than the others.

In the ENSEMBLES project, 15 institutes ran their RCMs at 25 km spatial resolution, with boundary conditions from five different GCMs, all using the same SRES emission scenario. In this study, it was decided to use one model per institute and only those models that extended their simulation until 2100. Table 39.2 lists the eight models that are used in this study (ENSEMBLES 2009).

RCM	Driving GCM	Reference
CNRM ALADIN	ARPEGE	[RAD 08]
DMI HIRHAM	ECHAM5	[CHR 06]
ICTP REGCM	ECHAM5	[PAL 07]
KNMI RACMO	ECHAM5	[VAN 08]
MPI REMO	ECHAM5	[JAC 01]
SMHI RCA	BCM	[KJE 05]
METOFFICE HadRM	HadCM3	[POP 07]
ETH CLM	HadCM3	[BÖH 06]

Table 39.2. *List of RCMs used in this study with their driving GCMs*

39.3.3. *Modeling frost damage*

The number of freeze-thaw cycles is calculated using the method according Ho andGough [HO 06], which counts if maximum temperature ≥ 0°C and minimum temperature ≤ −1°C. Pavement temperature is needed to calculate the number of freeze-thaw cycles. To calculate pavement temperature, the forecasted air temperature from climate models (daily minimum and maximum) is converted using the correlation equations derived from Dutch weather station and pavement sensor data. These equations are presented in Figure 39.4.

Based on analysis of the correlation between historic winter weather data and road damage, the threshold at which damage is expected to increase due to freeze-thaw cycles is 20 cycles per month. Each additional increase in five cycles will create further damage, thus thresholds are set at 20, 25, 30, and 35. If the number of freeze-thaw cycles increases beyond one of these thresholds during the lifespan of a road, the costs for both "adapt" and "no adapt" strategies are then calculated.

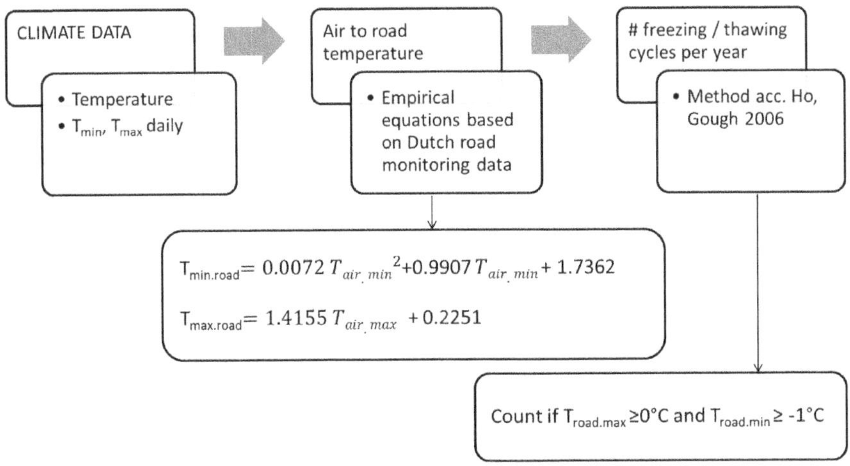

Figure 39.4. *Flow chart of the input climate related input parameters for predicting frost damage*

Adaptation requires that the road design is altered before being repaved. To redesign the road for increased levels of freeze–thaw, costs will be incurred. This is captured in the stressor cost increase variable, which is either calculated as the costs associated to modify pavement design (e.g. increase bitumen content, decrease aggregate size) or the costs of constructing a different pavement type, whichever is more appropriate given the anticipated threshold. General equations used for this calculation are presented below:

$$C_{\text{adapt}} = (N_{\text{thresh}} \times \text{SCI}) \times B_{\text{PG}} \qquad [39.1]$$

where C_{adapt} = increased cost due to climate change adaptation, N_{thresh}= number of thresholds crossed during road lifespan, SCI = stressor cost increase and B_{PG} = base construction cost = resurfacing costs.

Following the no adaptation strategy, the road design is not modified and damage will increase if thresholds are crossed. In this case, additional maintenance to repair the damage is required in order to preserve the original road design life. The increased costs are calculated by multiplying original winter maintenance costs by the expected percentage increase in damage, as calculated above during the threshold analysis. The advantage of this strategy is that the cost of additional maintenance may be less than the cost to modify the pavement design and construction. General methodology for calculating "no adapt" costs are presented below.

$$\mathrm{MT} = L_{\mathrm{ERB}} \times C_{\mathrm{ERB}} \times L_{\mathrm{road}} \qquad [39.2]$$

where MT = increased maintenance costs due to climate change, L_{ERB} = % change (decrease) in lifespan associated with a unit change in climate stress, C_{ERB} = current/base frost-related maintenance costs per km per year and L_{road} = length of roads affected by frost damage.

39.4. Conclusions

The combination of Rijkswaterstaat's reputation for long-term planning, strict maintenance procedures and recent budget constraints [RWS 12] further highlights the need for adaptation analysis that addresses their organization and regional priorities, not only sector-wide concerns. To obtain this, the research incorporates local empirical information, such as weather station and pavement sensor data. Combined with detailed climate models, pavement material properties and organizational cost data, the IPSS freeze–thaw modeling produces more accurate and actionable results than what is currently incorporated into existing decision-making. This improves support for design, maintenance and long-term planning within the organization.

It is particularly useful for the Netherlands to review the impacts of climate change in respect to analogous climate. The changing winter conditions and freeze–thaw patterns may be similar to environments regularly experienced by other countries, or by certain regions within the Netherlands. Taking this into consideration, the new design and maintenance strategies can be adjusted in advance to maximize the opportunity or minimize the impact. Unique to the Netherlands, however, is their reliance on PA. It is imperative for the long-term sustainability of their road infrastructure that PA roads will still be resilient and effective under future climate conditions.

As initial results from the freeze–thaw analysis in IPSS are being finalized, they support the thesis that in most regions it is not a question of if, but when adaptation will be required. The challenge that climate change presents for Dutch road infrastructure is both a short- and long-term concern. There are also potential opportunities from a changing climate in the Netherlands. In some regions, drier climate will require less robust drainage design and warmer winters could reduce freeze-thaw damage and subsequent maintenance on highways. The detailed level of analysis described in this chapter is meant to better inform the Dutch infrastructure operators about the regional variability, as well as long-term trends that threaten the performance and operating cost of the road network.

39.5. Acknowledgments

The project team would like to thank the Netherlands Organisation for Scientific Research (NWO) for funding through the Sustainable Accessibility of the Randstaad Project and Rijkswaterstaat, for their collaboration.

39.6. Bibliography

[ALV 06] ALVAREZ A.E., MARTIN A.E., ESTAKHRI C.K. *et al.*, "Synthesis of current practice on the design, construction, and maintenance of porous friction courses", Technical Report 0-5262-1, Transportation Research Board, 2006.

[BEN 05] BENDSTEN H., ANDERSON B., RAABERG J. *et al.*, "Two-layer porous asphalt – for urban roads." Report 144. Road Directorate, Danish Road Institute, Denmark, 2005.

[BÖH 06] BÖHM U, KÜCKEN M, AHRENS W. *et al.*, "CLM–the climate version of LM: brief description and long-term applications." *COSMO Newsletter 6*, 2006.

[CHI 11] CHINOWSKY P.S., HAYLES C., SCHWEIKERT A. *et al.*, "Climate change: comparative impact on developing and developed countries." *Engineering Project Organization Journal*, vol. 1, no. 1, pp. 67–80, 2011.

[CHI 12] CHINOWSKY P.S., ARNDT C., "Climate change and roads: a dynamic stressor–response model", *Review of Development Economics*, vol. 16, no. 3, pp. 448–462, 2012.

[CHI 13a] CHINOWSKY P.S., SCHWEIKERT A.E., STRZEPEK N.L. *et al.*, "Infrastructure and climate change: impacts and adaptations for the zambezi river valley", *UNU-WIDER, WIDER Working Paper*, 2013.

[CHI 13b] CHINOWSKY P.S., PRICE J.C., NEUMANN J.E. (ed.), "Assessment of climate change adaptation costs for the U.S. road network", *Global Environmental Change*, vol. 23, no. 4, pp. 764–773, 2013.

[CHR 06] CHRISTENSEN O.B., DREWS M, CHRISTENSEN J.H. *et al.*, "The HIRHAM regional climate model version 5 (β)", *DMI, Tech Rep.*, pp. 06–17, 2006.

[HO 06] HO E., GOUGH W.A., "Freeze thaw cycles in toronto, canada in a changing climate,", *Theoretical and Applied Climatology*, vol. 83, nos. 1–4, pp. 203–210, 2006.

[HUU 10] HUURMAN M., MO L., WOLDEKIDAN M.F., "Porous asphalt raveling in cold weather conditions", *International Journal of Pavement Research and Technology*, vol. 3, no. 3, 2010.

[IPC 07] IPCC, 2007: Summary for Policymakers. In: Climate Change 2007: The Physical Science Basis. Contribution of Working Group I to the Fourth Assessment Report of the Intergovernmental Panel on Climate Change, SOLOMON S., QIN D., MANNING M. *et al.* (eds.), Cambridge University Press, Cambridge, United Kingdom and New York, NY, USA, 2007.

[JAC 01] JACOB D., "A note to the simulation of the annual and inter-annual variability of the water budget over the baltic sea drainage basin", *Meteorol Atmos Phys.*, vol. 77, pp. 61–73, 2001.

[JAR 10] JAROSZWESKI D., CHAPMAN L., PETTS J., "Assessing the potential impact of climate change on transportation: the need for an interdisciplinary approach", *Journal of Transport Geography*, vol. 18, no. 2, pp. 331–335, 2010.

[JOT 10] JOTZO F., "Prerequisites and limits for economic modeling of climate change impacts and adaptation." *Environmental Economics Research Hub Research*, Report No. 55, The Crawford School of Economics and Government, Australian National University, Canberra Australia, 2010.

[KJE 05] KJELLSTRÖM E, BÄRRING L, GOLLVIK S, *et al.*, "A 140-year simulation of European climate with the new version of the rossby centre regional atmospheric climate model (RCA3)", *Rep Meteorol Climatol 108. Swedish Meteorological and Hydrological Institute*, Norrköping, 2005.

[KWI 13] KWIATKOWSKI K.P., STIPANOVIC OSLAKOVIC I., TER MAAT H.W. *et al.*, "Climate Change Adaptation and Roads: Dutch Case Study of Cost Impacts at the Organization Level", *Proc., Engineering Project an d Organization Conference,* EPOS, Winter Park, CO, 2013.

[MIL 09] MILLS B., TIGHE S., ANDREY J. *et al.*, "Climate change implications for flexible pavement design and performance in Southern Canada." *Journal of Transportation Engineering*, vol. 135, no. 10, pp. 773–782, 2009.

[MIR 09] MIRADI MARYAM., "Knowledge discovery and pavement performance: intelligent data mining", Delft University of Technology, The Netherlands, 2009.

[MOT 12] MOTE P., IDEKER J., FIGLIOZZI M. *et al.*, "Climate change impact assessment for surface transportation in the Pacific Northwest and Alaska." Technical Report, Washington State DOT, Washington State, U.S., p. 287, 2012.

[NEM 12] NEMRY F., DEMIREL H., "Impacts of climate change on transport: a focus on road and rail transport infrastructures", *European Commission, Joint Research Centre (JRC)*, Institute for Prospective Technological Studies (IPTS), 2012.

[PAL 07] PAL J.S., GIORGI F., BI X. *et al.*, "Regional climate modeling for impacts of future the developing world: the ICTP RegCM3 and RegCNET", *Bulletin of the American Meteorological Society*, vol. 88, no. 9, pp. 1395–1409, 2007.

[PET 08] PETERSON T.C., MCGUIRK M., HOUSTON T.G. *et al.*, Climate variability and change with implications for transportation, Transportation Research Board, Special report, 2008.

[POP 07] POPE V. *et al.*, "The Met Office Hadley Centre climate modelling capability: the competing requirements for improved resolution, complexity and dealing with uncertainty", *Phil. Trans. R. Soc. A.*, vol. 365, pp. 2635–2657, 2007.

[RAD 08] RADU R, DÉQUÉ M, SOMOT S., "Spectral nudging in a spectral regional climate model", *Tellus Ser A Dyn Meterol Oceanogr*, vol. 60, pp. 898–910, 2008.

[RWS 10] RWS, "Richtlijnen handelwijze bij vorstschade ann asphalt", *Rijkswaterstaat, Ministry of Infrastructure and the Environment*, The Netherlands, 2008.

[RWS 12] RWS, "Annual Review." *Rijkswaterstaat, Ministry of Infrastructure and the Environment*, The Netherlands, 2012.

[SU 13] SU Z., "A sustainable maintenance solution for porous asphalt pavements via rejuvenation technology", *Sustainable Construction Materials*, pp. 391–407, 2012.

[TOL 02] TOL R.S.J., "Estimates of the damage costs of climate change. part 1: benchmark estimates", *Environmental and Resource Economics*, vol. 21, no. 1, pp. 47–73, 2002.

[TRB 08] TRB, "Potential impacts of climate change on U.S. transportation", *Special Report, Transportation Research Board, National Research Council*, vol. 298, 2002.

[VAN 09] VAN DER LINDEN P., MITCHELL J.F.B. (eds.), ENSEMBLES: Climate Change and its Impacts: Summary of research and results from the ENSEMBLES project. Met Office Hadley Centre, FitzRoy Road, Exeter EX1 3PB, UK, pp. 10, 2009, available at http://ensembles-eu.metoffice.com/docs/Ensembles_final_report_Nov09.pdf

[VAN 02] VAN DER VALK A., "The dutch planning experience", *Landscape and Urban Planning*, vol. 58, nos. 2–4, pp. 201–210, 2002.

[VAN 05] VAN DER WAL J.S.I., DE BONDT A.H., "Case study life cycle cost analysis: ZOAB - maintenance of a highway", *Ooms Avenhorn Holding bv*, 2005.

[VAN 90] VAN DER ZWAN J., GOEMAN T., GRUIS H. *et al*, "Porous asphalt wearing courses in the netherlands: state of the art review", *Transportation Research Record 1265, TRB, National Research Council*, Washington D.C., pp. 95–110, 1990.

[VAN 08] VAN MEIJGAARD E, VAN ULFT L.H., VAN DE BERG W.J. *et al.*, The KNMI regional atmospheric climate model RACMO, version 2.1. TR-302, KNMI, De Built, 2008.

[VEJ 12] VEJDIREKTORATET, Winter service of porous asphalt European Experience, available at http://www.vejdirektoratet.dk/DA/viden_og_data/publikationer/Lists/Publikationer/Attachments/498/winter%20rapport.pdf, accessed February 2013.

40

Risk Assessment of Highway Flooding in the Netherlands

Based on results of the Risk Management for Roads in a Changing Climate (RIMAROCC) and SWAMP[1] projects in the ERANET Road call "Road owners getting to grips with climate change", the so-called "Blue Spots" study investigated the flooding events and the related likelihood of occurrence in detail for the Dutch highway network. Evidently, if risks need to be determined, the effects of occurrence of flooding events need to be determined as well. To assess the effects of the flooding events, five functionalities of the roads have been identified – and scored by experts – that are subject to change once a flooding event occurs. By describing likelihood and effects semi-quantitatively, the assessment of the risks can follow. Major results are discussed and steps further into the risk assessment are described, including the country-wide effects of failure of regional defenses in the water system and another study on the use of the road during evacuation and rescue, since flooding also affects the surrounding area. The chapter concludes with a positive statement on the applicability of the approach to other European countries.

40.1. Introduction

In Autumn 2012, Rijkswaterstaat, the Dutch National Road Authority, commissioned Deltares to establish along a semi-quantitative approach the risks at locations in the Netherlands National Highway Network sensitive to flooding due to changing climatic conditions. The rationale is that the risks and their relative importance will determine whether control measures will be necessary. If so, the first steps are establishing where, how and when measures can be taken.

Chapter written by Dirk PEREBOOM, Kees VAN MUISWINKEL and Thomas BLES.

1 Storm Water prevention – Methods to Predict damage from the water stream in and near road pavements in lowland areas

This chapter deals with the method of determining risks for the different types of highways in the national highway network and five distinguished flooding events with their respective frequency of occurrence. For the flooding events, only the effects on the flooded road and involved road network itself are examined. Possible measures are situated in or on the road network itself. The central question in this respect is: "How bad is it if requirements for the road and road network are not met, if a certain stretch of road is flooded?"

The study is a continuation of the study "Investigation of the Blue Spots in the Netherlands National Highway Network" by Bles *et al.* [BLE 12]. In this so-called "Blue Spots" study, the flooding events and the related likelihood of occurrence have been examined in detail for the highway network. Locations vulnerable to flooding on the Dutch highways were investigated, resulting in several maps and a report. Both the current situation and the situation in 2050 taking climate change into account were investigated. A logical next step was to investigate the related risks; in other words, is it a problem if a certain location on a highway floods? With that insight, Rijkswaterstaat will know the high-risk locations, can prioritize measures and choose locations that need further attention or investigation and detailed risk assessment.

40.2. Background – climate change and the mission of Rijkswaterstaat

The average global temperature is rising, as confirmed by IPCC Working Group I [IPC 13] in the first part of the fifth Assessment Report of IPCC. Sea level rise is expected to be more than previously estimated. According to KNMI, the Royal Netherlands Meteorological Institute, temperature in the Netherlands has risen more than the global average, and this trend is expected to continue. During the last 60 years precipitation has increased by 20%, and is expected to increase further. Besides that the frequency and intensity of extreme precipitation have increased, especially in winter, which, for example, means higher peak flows and levels in rivers. Sea level rise projections are near the global average, and little change in wind and storms is expected for the Netherlands.

Rijkswaterstaat is responsible for safe and fluid traffic on the National Highway network and to that end investigates the consequences of climate change for traffic. A climate resilient transport network will ensure that key corridors will be available to the user in all weather conditions. This is consistent with the vision on mobility presented by the European Road Transport Research Advisory Council (ERTRAC) in the European Roadmap Climate Resilient Road Transport. ERTRAC [ERT 11] states that an efficient and reliable transport system is essential for society while it is challenged by extreme weather events.

40.3. Development of methods to investigate vulnerability and to assess risks related to climate change

Before taking steps toward a climate resilient network of roads, vulnerable spots and the following risks have to be investigated. To that end, in 2007 ERA-NET ROAD made a call "Road owners getting to grips with climate change" running under the European Framework Project FP6[2]. Four projects were selected in 2007: IRWIN, P2R2C2, RIMAROCC and SWAMP. More information on these projects is found in the final report of ERA-NET ROAD program by Adesiyun *et al.* [ADE 11]. In this chapter, RIMAROCC and SWAMP methods relevant for the project are described, which have been implemented in the Netherlands.

SWAMP was implemented in the Netherlands by an assignment of Rijkswaterstaat, focusing on identifying the so-called blue spots (spots vulnerable to flooding). The work done resulted in the Deltares report "Investigation of the blue spots in the Netherlands National Highway Network" by Bles *et al.* [BLE 12]. The main results of the investigation are summarized in the following section 40.4.

In RIMAROCC, a method (see Figure 40.1) was developed for risk analysis and risk management with regard to climate change. The method was developed to support decision making concerning adaptation measures in the road infrastructure. In the previously mentioned assignment of Rijkswaterstaat, Deltares followed this method in the "Blue spots" study through to step 2.2, with vulnerability maps as a result. Within the risk assessment, the RIMAROCC framework has been followed further to step 4.3. Step 5.2, defining control measures, was to follow.

40.4. Summary of the "Blue spots" study

In the "Blue spots" study, flooding events and the related likelihood of occurrence have been investigated in detail for the highway network. In the survey, locations vulnerable to flooding on the Dutch highways were identified, resulting in several maps and a report. The report contains detailed information about the method and the results. In this chapter, the main results and conclusions are summarized as they are the basis for the risk assessment that is performed.

In the study, three main types of flooding were taken into account (see Table 40.1). Not every type of flooding has the same impact on the road infrastructure. The impact of flooding is very relevant when a risk assessment is performed, especially when measures are being identified and implemented. The

2 A subsequent call has been made in 2012 under the CEDR Transnational Road Research Programme: "Road owners adapting to climate change".

accent of measures for flooding with a high impact probably lies on evacuation and emergency planning, while measures for flooding with a low impact probably deal with traffic management. The impact of flooding is given in Table 40.1 as well.

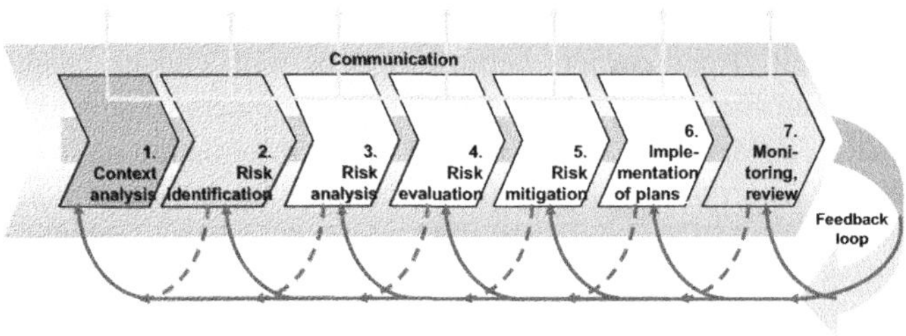

Key steps	Sub-steps
1. Context analysis	1.1 Establish a general context 1.2 Establish a specific context for a particular scale of analysis 1.3 Establish risk criteria and indicators adapted to each particular scale of analysis
2. Risk identification	2.1 Identify risk sources 2.2 Identify vulnerabilities 2.3 Identify possible consequences
3. Risk analysis	3.1 Establish risk chronology and scenarios 3.2 Determine the impact of risk 3.3 Evaluate occurrences 3.4 Provide a risk overview
4. Risk evaluation	4.1 Risk prioritisation 4.2 Compare climate risk to other kinds of risk 4.3 Determine which risks are acceptable
5. Risk mitigation	5.1 Identify options 5.2 Appraise options 5.3 Negotiation with funding agencies 5.4 Formulate an action plan
6. Implementation of action plans	6.1 Develop an action plan on each level of responsibility 6.2 Implement adaptation action plans
7. Monitor, re-plan and capitalise	7.1 Regular monitoring and review 7.2 Re-plan in the event of new data or a delay in implementation 7.3 Capitalisation on return of experience of both climatic events and progress of implementation
Communication and gathering of information	

Figure 40.1. *RIMAROCC risk management framework*

Type of flooding			Influence parameter	Physical effects	Impact	
					Duration	Area
A	Failure of flood defences	Flooding from sea and large rivers	Sea and river levels at certain frequencies	Flooding of the road Uplift / Heave Instability	Weeks to months	Dike ring areas (numerous square kilometers)
		Flooding from small rivers/canals			Days to weeks	Polder areas (several square kilometers)
B	Water system in the area around the road is not capable for drainage / discharge of water	Pluvial flooding (overland flow after precipitation)	Intense rainfall Long period of rain	Flooding of the road Uplift / Heave Instability	Hours to a day	Locally
		Increase of groundwater levels	Intense rainfall / long periods of rain	Uplift / Heave Instability	Weeks to months	Regional
		Increase of aquifer hydraulic heads	Intense rainfall / long periods of rain / sea level rise	Uplift / Heave	Weeks to months	Regional
C	Road surface not capable for enough drainage / discharge of water	Run-off on the road	Intense rainfall	Flooding (waterfilm) of the road	Minutes to hours	Locally
		Flooding of the storm water drainage system				

Table 40.1. *Three main types of flooding, effects and impact, from "Blue spots" study by Bles et al. [BLE 12]*

40.4.1. *Flooding due to failure of flood defenses*

For flooding type A, it is shown in the investigation that a large part of the Dutch highways can be affected to such an extent that it is no longer possible to use the road over large distances and over long periods of time. The probability of failure is low (regional defenses) to very low (primary defenses alongside sea and rivers).

40.4.2. *Flooding by intense rain and changing groundwater levels*

For flooding type B, the roads will only be locally affected, but still for longer periods. Some special objects can become less available or even unavailable. Although these are only local points on road trajectories, a whole trajectory still can become unavailable.

40.4.3. *Flooding by incapacity of storm water drainage and road surfaces*

The probability of flooding by incapacity of storm water drainage and road surface is relatively high and the number of blue spots can be more than doubled due to climate change. Blue spots are mostly present at locations with a change in transverse slope, many lanes and closed pavements.

For flooding type C, it is shown that the roads will be locally affected only for short periods of time. The effects of flooding type C are quite different from the effects of flooding types B and C. Flooding type C will not lead to a complete unavailability of the road, except for tunnels and deep lying sections.

40.5. Flooding events and their frequency of occurrence

In this section, the RIMAROCC steps 3.1 and 3.3 are elaborated, providing insight into the relevant scenarios that should be taken into account. Five types of flooding events affecting roads have been identified in the previous "Blue spots" study by Bles *et al.* [BLE 12]. These events showed high vulnerability on certain locations on the road network. The frequency of occurrence is also indicated:

1) more than 3 mm of water on the road surface during heavy rainfall (once per 10 years);

2) insufficient capacity in the storm water drainage system during heavy rainfall, resulting in water on the road surface either on the right or the left (traffic) lane (once per every 10 years, once per 250 years for tunnels);

3) an increase in groundwater levels due to climate change, causing possible uplifting of constructions such as tunnels, aqueducts, deepened roads and constructions with expanded polystyrene (EPS) blocks. In this study, we used the water levels that will occur every year to identify vulnerable locations. If these vulnerable locations learn to be high-risk locations, the design water levels of the constructions need to be specifically investigated in order to learn the water levels that are accounted for in the design. The probability of exceeding these water levels will be much lower than once per year and a new risk evaluation can be done afterwards;

4) insufficient bearing capacity or loss of stability of the road embankment due to flooding of the area caused by failure of primary defenses. The (temporary) effect of this event may be a higher groundwater table in the embankment. This will reduce geotechnical bearing capacity and/or embankment slope stability (once per 250 years to 10,000 years, depending on the location in the Netherlands);

5) complete flooding with more than 20 cm water on the road surface due to flooding of the area caused by failure of primary defenses (once per 250 years to 10,000 years, depending on the location).

40.6. Consequences of flooding

In the "Blue spots" study by Bles *et al.* [BLE 12], locations vulnerable to flooding on the Dutch highways were investigated, resulting in several maps and a report. Yet, if risks need to be determined, consequences or the effects of occurrence of flooding events need to be determined first. The consequences are different for more and less important road sections. This is taken into account in the investigation. This sections deals with the RIMAROCC step 2.3.

The Dutch highway network forms the connection between approximately 40 economical and governmental centers in the Netherlands and connects with the main infrastructure of surrounding countries. The network is coarse meshed and found in corridors, city ring roads and interurban connections. Around 2.2% of all paved roads in the Netherlands belong to the main network.

The roads of the network are classified in five different categories, four being Dutch categories:

– A – basic connections in highway network;

– B – connections between regional networks;

– C – connections between main economic destinations;

– D – ringway and links to main economic destinations;

and, as the fifth category, the Dutch parts of the Trans-European Transport Network (TEN-T) are added.

The division into the four Dutch categories A, B, C, and D is based on the combination of three criteria:

– function from a spatial and economic point of view;

– use based on (weighted) traffic intensity;

– robustness/vulnerability based on the probability of disruption and its effects (traffic disturbance, reserve capacity on the route and alternative routes).

Generally speaking, the importance of a road increases from category A–D. The importance of TEN-T connections is seen as minimally that of category C. The flooding events have an effect on the functionalities of the road. At Rijkswaterstaat, the functionalities are systematically described in terms of reliability, availability, maintainability and safety (RAMS) or RAMSSHEEP. RAMSSHEEP adds to RAMS extra functionalities: security, health, environment, economics and politics. For each of the abbreviated functionalities, further specifications are available.

To assess the effects of the flooding events, not all of the items of RAMS or RAMSSHEEP are useful. In consultation with experts of Rijkswaterstaat, five functionalities have been identified that are subject to change once a flooding event takes place. These functionalities (ASSER) are as follows:

– A: availability;

– S: safety;

– S: surroundings (effect on highway and other networks);

– E: euro;

– R: reputation.

40.7. Scoring effects of events

The scoring of the events is the elaboration of RIMAROCC step 3.2.

For every event (1–5) and every network category (A, B, C, D and TEN-T), six experts have indicated the influence of the events by scoring each of the five functionalities ASSER, irrespective of the location and related traffic intensity of a certain part of the network. Location and traffic intensity are already incorporated in the designation of roads into the A, B, C, D and TEN-T categories. So, respondents have been asked to merely judge the reduction in functionalities, irrespective of whether a specific part of the road is located in e.g. the central western part of the country, being more busy, or in a less frequently used part of the network more in the eastern or northern part of the country. The TEN-T category always coincides with one of the other categories A, B, C or D. In this case, every single stretch of road is given the highest score of the two.

Generally speaking, the larger the (negative) influence of the event on each of the functionalities of the road, the larger the score for the specific functionality, ranging from 1 to 4. For each of functionalities, further specifications are available. As an example in this chapter (see Table 40.2), the descriptions for availability and safety that have been given to the respondents are presented.

Score	Influence on availability	Influence on safety
1	Neglectably negative	Neglectable within accepted limits, light material damage only
2	Minimally negative	Close to accepted limits, increase in number of accidents, material damage, light injuries, no lasting health effects
3	Seriously negative	Beyond accepted limits, large increase in number of accidents, heavy material damage, heavy injuries, lasting health effects
4	Catastrophically negative	Far beyond accepted limits, extreme large increase in number of accidents, extreme heavy material damage, heavy injuries, deadly casualties

Table 40.2. *Descriptions for scoring effect (1–4) of flooding events for functionalities availability and safety*

40.8. Analysis and interpretation

During the period of analysis and interpretation (RIMAROCC step 3.4), scores have been discussed in an expert session. In the session, probabilities, resulting risks, and options for adaptation measures were also discussed.

The results on the requested indication of the relative importance of the functionalities to each other have also been discussed. As a result, the average relative ranking (1 = most important, 5 = least important) of the functionalities (see Table 40.3) has been determined.

In this chapter, some of the data have been summarized (see Table 40.4), presenting the number of respondents scoring 1, 2, 3 or 4, and the resulting average score (avg.). As can be seen, though the score is subject to personal opinions, background and experience, a large agreement resulted. It was concluded that, despite the limited number of respondents, the results are assumed to be representative in this semi-quantitative approach.

Functionality	Rank
Safety	1
Availability	2.2
Reputation	2.8
Surroundings	2.8
Euro	3.3

Table 40.3. *Average relative ranking (1 = most important, 5 = least important) of the functionalities of the road*

Flooding event	Network category	Score 1, 2, 3 or 4									
		Availability					*Safety*				
		1	2	3	4	Average	1	2	3	4	Average
(1) More than 3 mm of water	A	1	2	2	0	2.2	0	4	0	0	2
	B	1	2	2	0	2.2	0	4	0	0	2
	C	0	1	4	0	2.8	0	4	0	0	2
	D	0	1	4	0	2.8	0	4	0	0	2
	TEN-T	0	1	4	0	2.8	0	4	0	0	2
(2) Insufficient storm water drainage capacity	A	1	3	1	0	2	1	3	0	0	1.75
	B	1	3	1	0	2	1	3	0	0	1.75
	C	0	2	3	0	2.6	1	2	1	0	2
	D	0	2	3	0	2.6	1	2	1	0	2
	TEN-T	0	2	3	0	2.6	1	2	1	0	2
(3) Increase in groundwater levels	A	0	0	4	2	3.33	0	1	4	0	2.8
	B	0	0	4	2	3.33	0	1	4	0	2.8
	C	0	0	4	2	3.33	0	1	4	0	2.8
	D	0	0	2	4	3.67	0	1	3	1	3
	TEN-T	0	0	3	3	3.5	0	1	3	1	3
(4) Insufficient bearing capacity/loss of stability	A	0	0	4	2	3.33	1	2	2	0	2.2
	B	0	0	4	2	3.33	1	2	2	0	2.2
	C	0	0	3	3	3.5	1	2	1	1	2.4
	D	0	0	1	5	3.83	1	2	1	1	2.4
	TEN-T	0	0	3	3	3.5	1	2	1	1	2.4
(5) Complete flooding, more than 20 cm water	A	0	0	4	2	3.33	0	2	3	0	2.6
	B	0	0	4	2	3.33	0	2	3	0	2.6
	C	0	0	2	4	3.67	0	2	3	0	2.6
	D	0	0	2	4	3.67	0	2	3	0	2.6
	TEN-T	0	0	2	4	3.67	0	2	3	0	2.6

Table 40.4. *Scoring the effects of flooding events and network category on functionalities of the road. The larger the (negative) influence of the event on each of the functionalities, the larger the score (from 1 to 4) for that specific functionality*

In section 40.5 of this chapter, the frequencies of occurrence for the five types of flooding events have been presented. Depending on the event, the frequency of occurrence may vary from once per year to once per 10,000 years. To express these frequencies of occurrence also in a semi-quantitative probability scoring range from 1 to 4, the following set of formulations apply, transforming return period T into probability score P:

$$\text{Return period } 1 \text{ year} \leq T \leq 10 \text{ year} \quad P = 4 \qquad [40.1]$$

$$\text{For return period } T > 10 \text{ year} \quad P = 5 - \log(T) \qquad [40.2]$$

This formulation adds the highest probability score P to the most frequently occurring events. For all events with a frequency of occurrence of less than once per 10 years, P diminishes from 4 to ultimately 1 for the least frequent event of once per 10,000 years.

Risk is defined as the multiplication of probability and effect. Both scores on probability and effect range from 1 to 4. Numerically, the score on risk varies between minimally 1 and maximally 16.

The scores of the individual respondents as "integer" 1, 2, 3 or 4 have been averaged per functionality, leading to a "real" outcome in the range of 1 to 4. The scores on probability also range from 1 to 4. The scores of the resulting risk range from 1 to 16. To translate these values into graphical representations, classes of effect, probability and risk have been defined (see Table 40.5).

40.9. Risk evaluation (RIMAROCC step 4)

Here, the discussion of the results is limited to safety and availability. Most emphasis is given to availability. Though second in ranking of importance, the scores on availability are the most prominent ones.

Presenting all the results in this chapter does not lead to a better understanding of the approach followed or its results. Hence, only the results on availability are presented graphically, limited to events 1 and 4, bearing in mind that the results for event 2 are comparable to those for event 1, and results for event 5 are comparable to those for event 4. Results for event 3 will also be shown. For this event, the results are large in value of risk. The number of sites is limited to approximately 20 well-defined locations.

40.9.1. *Results on availability, flooding events 1 and 4*

For flooding event 1, the network categories A and B score on effect on availability (see Figure 40.2(a)) less than 2.5, with designation moderate and yellow color, whereas for network categories C, D and TEN-T, the score is higher than 2.5 but below 3.25, with designation large and orange color.

Flooding event 1 has a high score on probability (4, the event has a frequency of occurrence of once per 10 years). The multiplying effect and probability give the risk score for all network categories (see Figure 40.2(b)) of higher than 8.5 and less than 12.25. This results in designation large and orange color.

For flooding event 4, all network categories score high on effect on availability (see Figure 40.3(a)), being higher than 3.25, with designation very large and red color.

Score with respect to		Designation	Colour
Effect - Probability	Risk		
1 – ≤ 1.75	1 – ≤ 4.75	Small	Green
1.75 – 2.5	4.75 – 8.5	Moderate	Yellow
2.5 – 3.25	8.5 – 12.25	Large	Orange
≥ 3.25 – 4	≥ 12.25 – 16	Very large	Red

Table 40.5. *Classes of effect, probability and risk used in graphical representations*

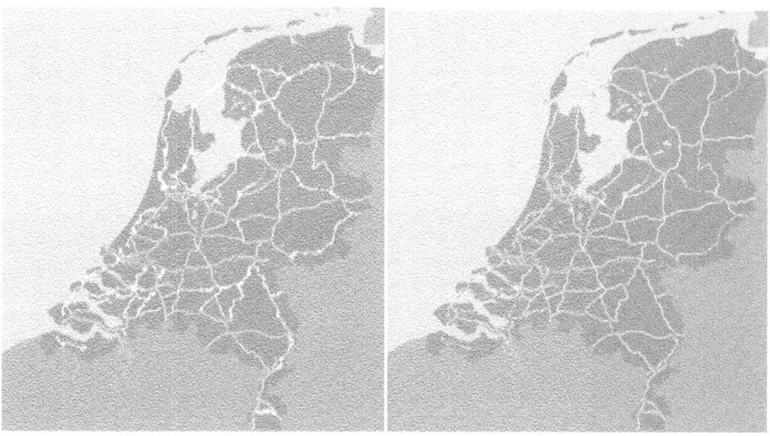

a) b)

Figure 40.2. *Flooding event 1 (only for functionality availability): a) effect score and b) risk score. For a color version of this figure, see www.iste.co.uk/jacob/materials.zip*

For flooding event 4, these high scores on effect are multiplied with low values on probability. The events occur in areas with frequencies of occurrence of less than once in 250 years or smaller, down to as small as once in 10,000 years. This results in rather low scores for risk (see Figure 40.3(b)).

Only for three locations near the River Meuse in the south east of the country, an effect score of 3.5 for the locally present network category C, or TEN-T, multiplied with a score on probability of 2.6 (once in 250 years), yields a score on risk of 9.1, higher than 8.5 and less than 12.25, with designation large and orange color.

For all other locations, even the highest effect score of 3.8 for network category D is maximally multiplied with a score on probability of 1.9 (once in 1,250 years) with designation moderate and yellow color. Or less, because for frequencies of occurrence below once in 1,250 years, and for all effect scores, that vary between 3.3 and 3.8, the resulting scores on risk are less than 1.9, down to 1. This results in designation small and green color, for large parts of the Netherlands.

For some locations, only scores on effects and no scores on risks are visible. In the modeling of the flooding events, some inundation is calculated for these locations as well. This means the locations fall into category 5. However, since these locations are situated outside an area surrounded by embankments, they do not have a value for frequencies of occurrence related to the design water levels of the defense structures. For practical reasons, the effects are shown on the maps, whereas multiplication of effect with probability does not have a meaning here. Therefore, results on risks are not shown for these locations.

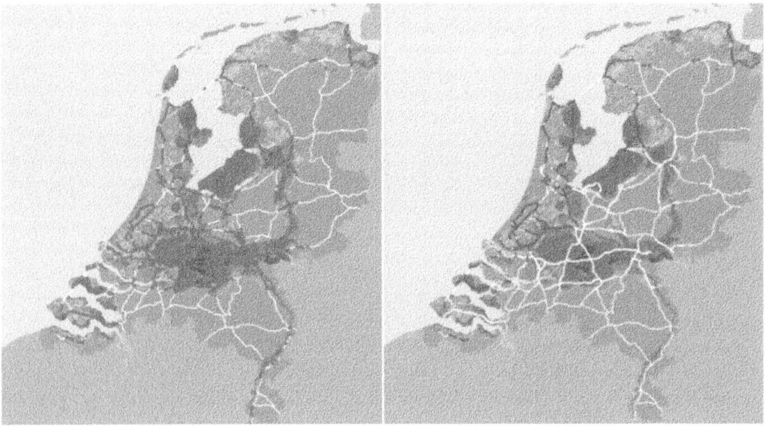

a) b)

Figure 40.3. *Flooding event 4 (only for availability): a) effect score and b) risk score. For a color version of this figure, see www.iste.co.uk/jacob/materials.zip*

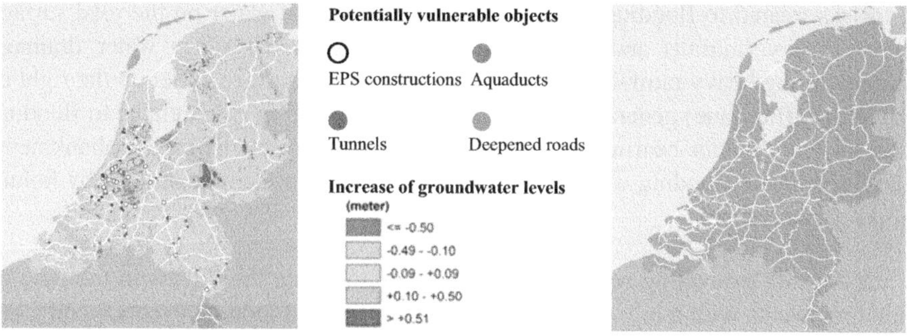

Figure 40.4. *Flooding event 3 (only for availability): (a and b) potentially vulnerable objects; increase in ground water levels; (c) risk score of approximately 20 sensitive locations. For a color version of this figure, see www.iste.co.uk/jacob/materials.zip*

40.9.2. *Results on availability, flooding event 3: increase in groundwater levels due to climate change*

For flooding event 3 (see Figures 4.4(a) and (b)), an increase in groundwater levels and potentially vulnerable locations are combined. High scores on effects on availability are multiplied with a high score on probability (4, the frequency of occurrence is once per 10 years). As a result, scores on risk are higher than 12.25. These scores (see Figure 4.4(c)) have designation very large and red color.

Potentially vulnerable objects are restricted to locations where EPS has been used and locations with constructions such as tunnels, deepened roads and aqueducts, in combination with locations where the groundwater level will rise in the coming years until 2050 due to climate change. The number is limited to approximately 20 sensitive locations. This number is small, because for many locations in the Polder areas in the western and northern part of the country there is no chance for lifting up. In all these areas with an artificially maintained surface water level, it is assumed that groundwater level will not change compared to the surface level and the level of the roads. A possible lowering of groundwater level will be accompanied by the same change in surface level and the level of the road. Therefore, from many locations, only a limited number finally appears on the risk map.

40.10. Conclusions

Flooding event 3 (an increase in groundwater level causing uplifting of constructions), though only on a limited amount of sensitive locations, generates the highest risk score.

Risks related to flooding event 1 (more than 3 mm of water on the road surface during heavy rainfall) and 2 (insufficient capacity in the storm water drainage system during heavy rainfall, causing water on the road surface either on the right or the left (traffic) lane) generate a higher score or are equal to risks related to flooding event 4 (insufficient bearing capacity or loss of stability of the road embankment) and 5 (complete flooding with more than 20 cm water on the road surface by failure of primary defense).

Alongside evaluating risks, it is necessary to evaluate the effects in the case of flooding events with a low probability and large effects. Flooding events 4 and 5 are such events with large effects.

Against expectations, apart from locations where uplifting of constructions by an increase in groundwater levels is possible (event 3), no other specific possible problem locations have been found by assessing risks. To be more commonly expected is the unavailability of complete road trajectories.

40.11. Recommendations

One important recommendation is to further investigate special locations such as engineering works, traffic interchanges and key tunnels in the Rotterdam harbor area, potentially causing large-scale traffic disruption when flooded. Also the locations that are potentially sensitive to an increase in groundwater levels (approximately 20 locations, of which about 50% have a very large risk) have to be studied in more detail.

Alongside evaluating risks, it is necessary to evaluate the effects in the case of flooding events with a low probability and large effects. Flooding events 4 and 5 are events with large effects.

Results reported later in 2013 from another study about flooding affecting the surrounding area need to be integrated in the present results. This is expected to give further indications of how and where priorities are to be addressed. The impacts of failure of regional defenses should also be taken into account. Studies finalizing the probabilities of failure of regional defenses for the Netherlands as a whole are now available, so implementation of these findings can follow.

40.12. Implementation of results in the Netherlands and other European countries

Following the above recommendations, further studies and implementation of results remain. Assessing the risks of regional defenses needs to be done. The availability of roads during evacuation and rescue also needs to be addressed, since flooding also affects the surrounding area. These studies may provide more indications on setting priorities in terms of where and when there is the largest need of control. This is one step further in the process, to be followed by RIMAROCC step 5, risk mitigation itself.

During the study, the well-known difference between "risk management" in general, and performing "the real steps to a substantial result" in detail, has been encountered, as per usual in risk assessments. Although "risk management" consists of relatively easily explainable steps (seven steps when using RIMAROCC), "the real steps to a sustainable result" require an elaborate and detailed analysis if at least semi-quantitative results are to be attained. The "Blue spots" study, being the starting point for the risk analysis, dealt with events and probabilities requiring analyses and interpretation on many detailed aspects. Almost all required knowledge and data are present among the experts; nevertheless, it needs to be mobilized. The same challenge applies to researching the effects of flooding events presented in this chapter. The method used, consulting experts, was welcomed positively on their side and proved to be effective after a thorough preparation. In presenting the results, the GIS environment has been indispensable, again requiring detailed processing of all separate data on network categories, flooding events, functionalities, effects, probabilities and, finally, risks.

The application of the approach used here should also work in other European countries, using country-specific knowledge and data. This may lead to country-specific adaptation of the approach. To conclude, the authors would recommend everybody to do so, starting with a more global coverage of the complete network, to be refined later where necessary.

40.13. Acknowledgments

The authors wish to thank Rijkswaterstaat employees who have provided data and expert views on effects and risks of flooding.

40.14. Bibliography

[ADE 11] ADESIYUN A. *et al.*, "Road owners getting to grips with climate change", (final report of ERA-NET ROAD Programme), available at www.eranetroad.org, March 2011.

[BLE 12] BLES T.J. *et al.*, "Investigation of the blue spots in the Netherlands National Highway Network", Deltares report 1205568-000-GEO-0007, Version 2, available at www.fehrl.org, 15 May 2012.

[ERT 11] ERTRAC, "European roadmap climate resilient road transport", ERTRAC Research and Innovation Roadmaps, available at www.ertrac.org, pp. 182–198, September 2011.

[IPC 13] IPCC WORKING GROUP I, "Working Group I contribution to the IPCC Fifth Assessment Report – Climate Change 2013: The physical science basis, summary for policymakers", available at www.ipcc.ch, 2013.

41

Adaptation of the Road Infrastructure to Climate Change

Various climate projections predict changing climatic parameters such as temperature, precipitation and wind speed for Germany. This could have severe impacts on road transport infrastructure as well as road traffic itself. At the Federal Highway Research Institute (Bundesanstalt für Straßenwesen), a strategy was developed to adapt roads and engineering structures to the impacts of climate change. The strategy "*Anpassung der Straßenverkehrsinfrastruktur an den Klimawandel*/Adaptation of road traffic infrastructure to climate change (AdSVIS)" currently comprises about 15 projects. On the basis of the identification of the hazards and the combination of the climate and road network data, the road transport infrastructure that might be affected is to be determined. Adaptation measures are to be developed for the identified risk areas and consequently their effectiveness has to be assessed. Special attention is given to international cooperations since climate change is a truly global challenge.

41.1. Introduction

Road transport infrastructure has a huge economic importance. Currently, about 70% of goods are carried by road and the trend is still rising [ICK 07]. The road transport infrastructure is facing many challenges and the subsequent adaptation of the infrastructure is of utmost concern. These challenges are as follows: globalization, sustainability, technological and demographic change, an increase in goods transport and climate change.

Contrary to popular belief, the climate of our planet was never stable. But in recent years, the average global temperature has been rising quite significantly. Due

Chapter written by Markus AUERBACH and Carina HERRMANN.

to the fact that the CO_2 concentration in the atmosphere continues to increase, there will be a further rise in temperature [IPC 07]. Therefore, there is the immediate need to investigate how the climate will change, how this change will impact road transport infrastructure and how we can possibly adapt to that change.

To answer the first part of the question, we have to make use of emission scenarios and climate models. In 2007, the Intergovernmental Panel on Climate Change (IPCC) published its Fourth Assessment Report [IPC 07], where different emission scenarios were described. These scenarios represent different assumptions for the development of the world, e.g. the rate of increase in population, the rate of economic growth and the use of fossil or non-fossil energy sources. These scenarios are integrated into climate models and have been frequently used to make projections of future climate change. For Germany, an increase in extreme weather events, such as periods of heat and heavy rain events, is predicted.

41.2. Strategies of adaptation to climate change

41.2.1. *The German adaptation strategy*

As a reaction to climate warming, the Federal Government of Germany passed the "German Adaptation Strategy to Climate Change (DAS)" [BUN 08] in 2008. This strategy presents a framework for adaptation to the impacts of climate change in Germany. The aim of the strategy is to create a national framework for action in order to prevent or minimize the adverse effects of climate change on the public, natural habitats and the national economy. This framework is intended to make it easier to identify impacts and adaptation needs to plan and implement measures. For instance, early incorporation of adaptation aspects into planning can save costs associated with climate change in the future.

Besides outlining the current status of knowledge on the anticipated changes or climate and the impacts these could entail, the strategy also highlights possible climate impacts and options for action for 14 sectors and selected regions.

In the sector of road transport/road transport infrastructure, there are two key aspects for the medium term need for actions:

– adaptation of the dimensions of the drainage infrastructure to more extreme rainfall;

– adaptation of pavements affected by heat through the use of modified materials.

On this basis, the government adopted the "Action Plan to the German Adaptation Strategy to Climate Change" [BUN 11] in 2011. One objective of this action plan is to develop a set of indicators for the German Strategy for Adaptation to Climate Change. In the action plan, the climate impacts and individual adaptation methods in the sectors covered by the DAS are described. By using existing data, changes in each sector are documented and results for public activities are obtained.

In the road transport/road transport infrastructure sector, the Federal Highway Research Institute (Bundesanstalt für Straßenwesen, BASt) was involved in the development of three indicators [SCH 11]:

– weather-related road accidents;

– road condition;

– use of de-icing salt on federal roads and highways.

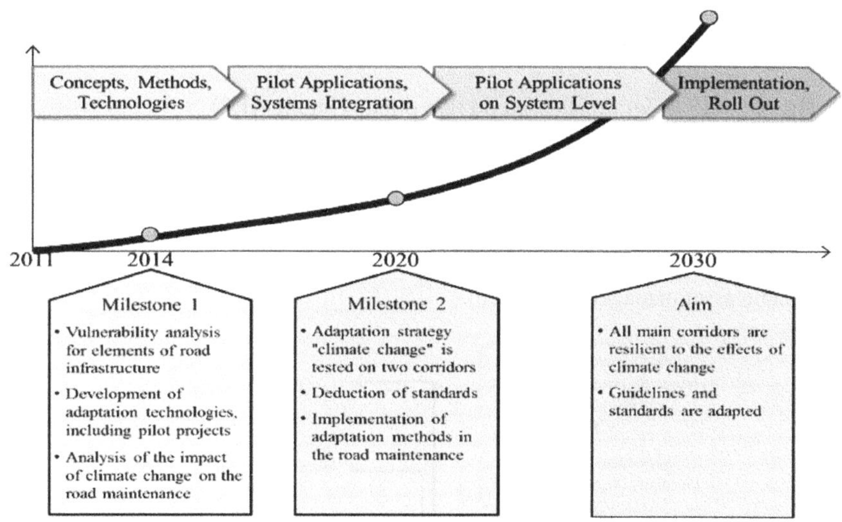

Figure 41.1. *Milestones of the BASt roadmap to the adaptation of the road infrastructure to climate change*

41.2.2. *The strategy of the Federal Highway Research Institute*

The German federal road network consists of more than 53,000 km of roads and more than 38,000 bridges and tunnels. The expected service life of roads is 30–50 years and for bridges 80–100 years (most bridges in Germany were built between 1965 und 1985). With respect to this, we already have to adapt the guidelines and standards to the impacts of the predicted climate change. The Federal Highway

Research Institute developed a roadmap (Figure 41.1) on how to reach this ambitious goal.

By 2014, a vulnerability analysis for approximately 1,200 km of the German roads, a few bridges and tunnels will be performed and the first adaptation technologies will be developed and tested in pilot projects. Besides, the impacts of climate change on road maintenance, especially the winter services, are analyzed. For Milestone 2, the adaptation strategy "climate change" will be tested on two corridors, standards will be defined and the adaptation methods for road maintenance implemented. Finally by 2030, it is expected that all main corridors will be resilient to the effects of climate change and the guidelines and standards will be adapted.

The implementation of this roadmap is a very complex task. To reach Milestone 1, the Federal Highway Research Institute initiated the program called "Adaptation of the Road Infrastructure to Climate Change (AdSVIS)" [TEG 11], which currently consists of 15 projects.

The AdSVIS program is based on two pillars (Figure 41.2):

– the metrological/geographical pillar;

– the adaptation pillar.

The aim of this study is to integrate this adaptation strategy into a holistic and sustainable asset management system.

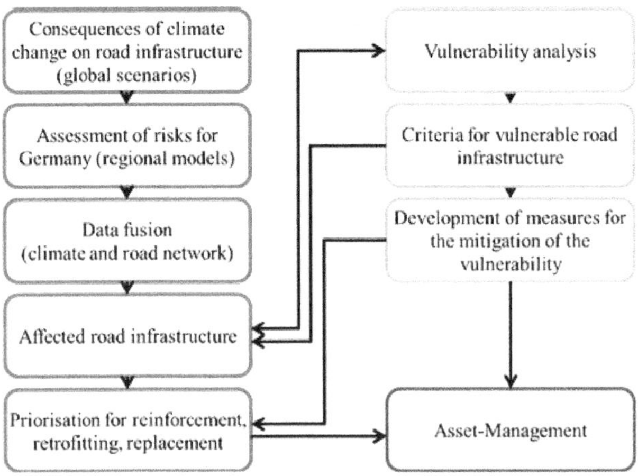

Figure 41.2. *Structure of "AdSVIS – adaptation of the road infrastructure to climate change"*

On the basis of the identification of the threats and the fusion of the climate and road network data, the road transport infrastructure that might be affected is to be determined. For this purpose, it is necessary to carry out initial vulnerability analyses and to establish the criteria for the elements of the road transport infrastructure which are at risk. At the same time, potential adaptation measures will be developed and their effectiveness will be assessed in a holistic approach. Subsequently, a prioritization for the implementation of the adaptation measures could be performed for the entire network. Each of the 15 AdSVIS projects has been assigned to one of the pillars. The projects are listed in Table 41.1 and are described in detail below.

Projects	*Status*
Cross-linking of the AdSVIS projects	
AdSVIS server	
RIVA - Risk analysis of key goods and transit axes including seaports	
Comparison of meteorological parameters near federal roads and grid data of climatologic prediction models	
Development of climate impact models and design parameters for bridges and tunnels	
Vulnerability analysis of bridges and tunnels	
Analysis of measures to reduce the vulnerability of bridge and tunnel structures	
Development of a model for estimating landslide risk areas and development of a national hazard map	
Evaluation of the design of road drainage systems regarding climate change	
Inventory of road drainage systems in selected sections of the TEN-T	
Adaptation of pavement designs for asphalt and concrete constructions	
Standardized asphalt pavements due to changing temperature boundary conditions	
Asphalt pavements and extreme temperatures	
Impacts of weather extremes on concrete road pavements	
Impacts of climate change on road maintenance	
Project running / *Project planned*	

Table 41.1. *The status of the AdSVIS projects. Running: light gray; planned: dark gray*

41.3. The AdSVIS projects

41.3.1. *The RIVA project*

The most important project of AdSVIS is "RIVA – risk analysis of key goods and transit axes including seaports" [ALF 12, ALF 13]. The aim of the project is the development of tools for the identification, analysis and assessment of risks due to the expected climate change.

The RIVA project ties in with the ERA-NET ROAD project RIMAROCC [BLE 10], in which conceptual basics for risks of the road infrastructure were considered (Figure 41.3). However, this methodology had to be adapted to the complexity of the German road network. For this purpose, an MS Excel based method, using standardized measurement data (e.g. performance data of the roads) and data of climate projections, was developed. This method was applied on nine sections making up about 10% of the German Federal Highway Network.

The results received using the RIVA method allow a first prioritization of climate-related risks. But they can also be used for a comparison with other road infrastructure risks. Therefore, the RIVA project provides an important contribution to the discussion of risk management strategies as well as to decision making on necessary measures.

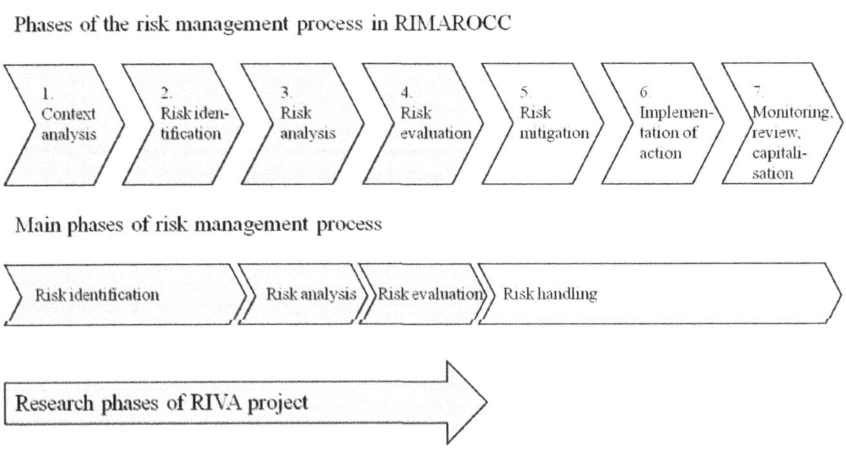

Figure 41.3. *The RIVA method (Source: RIMAROCC/Alfen Consult)*

41.3.2. *Comparison of meteorological parameters and grid data of climatologic prediction models*

Weather stations are rarely located close to roads and climate projections provide data for a specific grid. In these grids, the average values of an area are independent of topography or land use. In this project, algorithms, which compare the data of weather stations (temperature, precipitation) with weather data at roads, are created. Different types of construction and the location of the roads are considered. In the next step, the procedure is applied to projected climate data. Thus, the effects of climate change (e.g. rising temperatures) on road constructions are estimated. With this information, adaptation measures are developed and implemented.

41.3.3. *Cross-linking of the AdSVIS projects*

All AdSVIS projects are cross-linked with each other and the results and interim results are exchanged during the lifetime of each project. For example, the areas prone to future landslides (from the project "development of a model for estimating landslide risk areas and development of a national hazard map" (Table 41.1)) could be integrated in the risk analysis of the RIVA project. Furthermore, the results of AdSVIS will be presented at national and international events.

41.3.4. *AdSVIS server*

Almost all AdSVIS projects require network data of road infrastructure and climate projection data. The network data and data for civil engineering structures are contained in various databases. To simplify the handling of the different databases and to provide a consistent database to all users, the establishment of an all-encompassing database is planned. Furthermore, climate projection data of different climate scenarios and a digital terrain model will be integrated. In this way, all projects will have access to consistent data and the results will be comparable to each other.

41.3.5. *Development of climate impact models and design parameters for bridges and tunnels*

One of the results of the pilot project "Impacts of climate change on existing prestressed concrete bridges" [NOV 12] was that adaptations are necessary in particular for frame bridges made up of prestressed concrete which were built before 1981.

This project is aimed at determining site-related and design-relevant climate parameters in order to obtain projections of climate impacts which are as realistic as possible for structures that are located in regions particularly affected by climate change.

41.3.6. *Vulnerability analysis for bridges and tunnels*

The aim of this project is to identify critical structures so that the relevant measures can be considered to reduce the vulnerability of the structure. Thus, it may be possible to reduce the cost of repair or reconstruction of a damaged or destroyed structure.

41.3.7. *Analysis of measures to reduce the vulnerability of bridge and tunnel structures*

This project is based on the results of the project "Vulnerability analysis of bridges and tunnels". It will develop a catalog of measures to reduce the vulnerability of bridges and tunnels. The analysis will consider construction and organizational provisions of existing engineering structures as well as for future engineering structures. This catalog should serve as the guideline document for road authorities, and help to design and build new structures as well as reconstruct and adapt existing ones.

41.3.8. *Development of a model for estimating landslide risk areas and development of a national hazard map*

In the pilot project "Assessment of the risks of landslides by the increase in extreme weather events" [KUM 12], past landslide events have been analyzed and evaluated for three regional cases. In this project, landslide events are studied nationwide for Germany and, in addition to the slope inclination, which was the only variable evaluated in the pilot project, other parameters such as geology, geomorphology, groundwater, exposure and vegetation are considered. In combination with the climatic parameters, a climate engineering geological model for estimated landslide risk areas and a national hazard map for landslide risk areas along roads will be developed.

41.3.9. *Evaluation of the design of road drainage systems regarding climate change*

This project is intended to review the current design concepts in accordance with the guidelines for the design of roads (regarding drainage (RAS-Ew)) and the guidelines for structural measures at roads in water protection zones (RiStWag) with regard to the hydraulic efficiency while observing soil and water protection and the structural designs. The results of this research project are to serve as a basis for the revision of the regulations for new construction as well as maintenance.

41.3.10. *Inventory of road drainage systems in selected sections of the TEN-T*

In the context of this project, information of road drainage systems (open/closed) and retention basins will be prepared and implemented in the AdSVIS server. This inventory could also be used by the RIVA project.

41.3.11. *Adaptation of pavement design for asphalt and concrete constructions*

The methods currently used to take into account weather conditions for the design of road pavements (in the case of new construction as well as maintenance) are based on long-term series of meteorological observations. This project investigates the extent to which climate change modifies the input parameters which are subject to weather conditions. This refers to those parameters that are integrated into the guidelines for the standardization of pavements of traffic areas (RStO) as well as into the computational design (guidelines RDO asphalt[1] and RDO concrete[2]). It is intended to prepare the relevant adaptation proposals for design (30 years as a rule). In this project, the predicted increase in traffic load was not taken into account.

41.3.12. *Standardized asphalt pavements due to changing temperature boundary conditions*

The properties of asphalt are temperature dependent. Its viscoelastoplastic behavior causes asphalt to be prone to cracking in winter and permanent deformation (rutting) in summer. On the basis of the guidelines for the computational design of the surface layer of the asphalt pavement (RDO asphalt[1]), it is to be examined to what extent a rise in the average annual temperature and the impact of traffic allow the use of standard asphalt design and construction methods in order to reach the usual service life. Failure to reach the full service life will lead

1 Richtlinien für die rechnerische Dimensionierung des Oberbaus von Verkehrsflächen mit Asphaltdeckschicht.

2 Richtlinien für die rechnerische Dimensionierung von Betondecken im Oberbau von Verkehrsflächen.

to the development of new standards for material adaptation or the use of alternative binders.

41.3.13. *Asphalt pavements and extreme temperatures*

The structural capacity of asphalt pavements is achieved by changing the layer thickness or by changing the material properties. The thermophysical variation of the properties of the asphalt has been neglected. In this project, asphalt pavements should be adapted to climate change, especially to higher temperatures, by changing the asphalt type or the mineral aggregate.

41.3.14. *Impacts of weather extremes on concrete road pavements*

The aim of the project is to investigate the impact of various weather extremes on concrete pavements and to analyze if service lifetimes are reduced. With the study, critical or sensitive design is to be identified and repaired. The project is necessary because a sudden collapse of the road construction, such as a "blow up", is very dangerous.

41.3.15. *Impacts of climate change on road maintenance*

To determine the impact of climate change on road maintenance, climate events have to be forecasted. Hence, attention is paid to all services like snow and ice removal, de-icing, mowing of grass, pruning of brushes, cleaning of roads, cleaning and maintenance of drainage systems, removing of storm damage, etc. In addition, road traffic and traffic delays resulting from maintenance activities contribute to climate change. Particularly, traffic congestion produces pollution such as the generation of additional greenhouse gases that contribute to climate change. The aim of the project is to enhance maintenance services and reduce the influences and consequences of climate change.

41.4. International cooperation

Since climate change is a global challenge, international cooperation as an approach to tackle related problems is appealing, even more because the vulnerable infrastructure is not confined to a single country. BASt is therefore part of several international programs, often in a leading role.

41.4.1. *Forever open road*

The "Forum of European National Highway Research Laboratories (FEHRL)" started its flagship program "Forever Open Road (FOR)" in 2011, trying to redefine road transport in the 21st Century. One of three elements of this program is "the climate change resilient road", led by BASt (Figure 41.4).

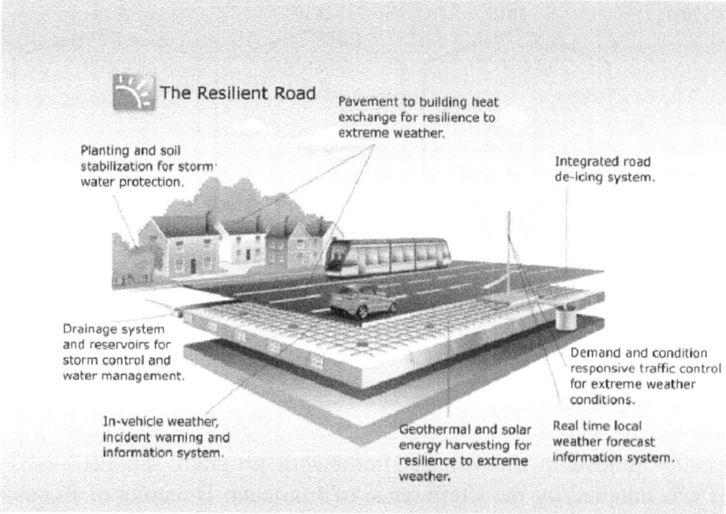

Figure 41.4. *The "resilient road" element of forever open road (FEHRL)*

The resilient road will adapt itself to the impacts of extreme weather conditions and climate change. The road will monitor flooding, snow, ice, wind and temperature change, and mitigate their impacts through integrated storm drainage, automatic heating and cooling, and will be linked to the integrated information system for travellers and operators.

The climate change resilient road focuses on ensuring adequate service levels of the road network under extreme weather conditions. Innovation themes will address adaptation of road operations and management to the effects of extreme weather to such an extent that adequate service levels are ensured.

All projects of the national FEHRL institutes, contributing to the aims of FOR that were awarded the FOR label, are listed in the FOR-portfolio [FOR 12]. The FOR strategy was published in three roadmaps [FOR 12]. The roadmap of the "resilient" element sets milestones that are in line with the German AdSVIS but focused more on the European level with its TEN-T network (Figure 41.5).

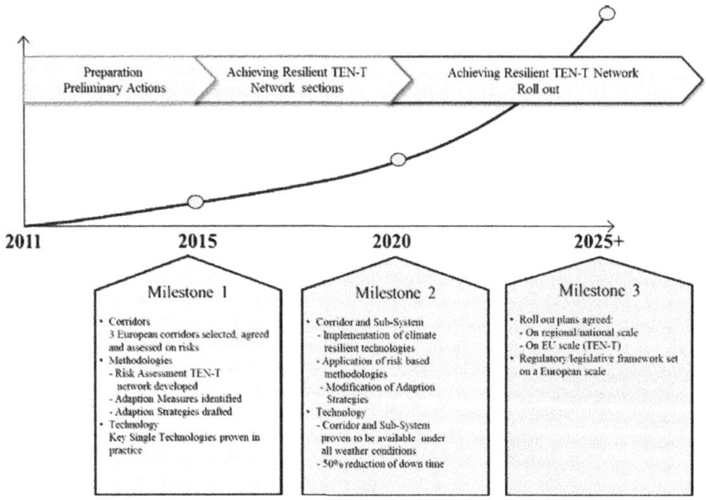

Figure 41.5. *Milestones of the FOR roadmap "the climate change resilient road"*

41.4.2. *CEDR call 2012*

Originally funded in the 7th EU framework program, the ERA-NET ROAD program was adopted by the Conference of European Directors of Roads (CEDR) and continued in 2012 among others with a call "Road Owners Adapting to Climate Change". Two projects were chosen: "Roads for today, adapted for tomorrow (ROADAPT)", a follow up of the "RIMAROCC" project with the aim of further developing this methodology based on user's experiences. The second project "Climate projection data base for roads (CliPDaR)" produces guidelines for a harmonization of meteorological input data in order to guarantee an at least Europe-wide comparability of climate data and their consecutive vulnerability assessments.

41.5. Conclusions

As a result of climate change, extreme weather events such as storms, heavy rainfall and heat waves will increase and the road transport infrastructure has to adapt accordingly. Thus the Federal Government of Germany passed the "German Adaptation Strategy to Climate Change (DAS)" and the "Action Plan to the German Adaptation Strategy to Climate Change". The Federal Highway Research Institute was involved in the development of three indicators (weather-related road accidents, road condition, use of de-icing salt on federal roads and highways) in the road transport/road transport infrastructure sector.

In 2009, the Federal Highway Research Institute (BASt) initiated a series of research projects related to climate change and developed an adaptation strategy. This strategy was implemented through a research program called "Adaptation of the Road Infrastructure to Climate Change (AdSVIS)". Currently, AdSVIS consists of 15 projects, of which the RIVA project is the most important project. The aim of RIVA is the development of tools for the determination of the need for action due to the anticipated climate change. In this way, risks for the infrastructure could be reduced and mitigated. Other projects of the AdSVIS program concern the adaptation of guidelines to climate change, in particular the adaption of pavements and the design of drainage systems. Nine projects have already been started and more are planned.

The Federal Highway Research Institute (BASt) in a leading role is also involved in several international programs such as "The Climate Change Resilient Road" in the FOR program or in the CEDR call 2012 "*Climate Change* – Road Owners Adapting to Climate Change".

41.6. Bibliography

[ALF 12] ALFEN CONSULT, Risikoanalyse wichtiger Güter- und Transitverkehrsachsen unter Einbeziehung von Seehäfen, 1. Interim report (in German), Bundesanstalt für Straßenwesen, 2012.

[ALF 13] ALFEN CONSULT, Risikoanalyse wichtiger Güter- und Transitverkehrsachsen unter Einbeziehung von Seehäfen, 2. Interim report (in German), Bundesanstalt für Straßenwesen, 2013.

[BLE 10] BLES T., ENNESSER Y., FADEUILHE J.-J. *et al.*, Risk Management for Roads in a Changing Climate - A Guidebook to the RIMAROCC Method, ERA-NET ROAD, 2010.

[BUN 08] BUNDESREGIERUNG, Deutsche Anpassungsstrategie an den Klimawandel, available at www.bmu.de/fileadmin/bmu-import/files/pdfs/allgemein/application/das_gesamt_bf.pdf, 2008.

[BUN 11] BUNDESREGIERUNG, Aktionsplan Anpassung der Deutschen Anpassungsstrategie an den Klimawandel, available at http://www.bmu.de/fileadmin/bmu-import/files/pdfs/allgemein/application/pdf/aktionsplan_anpassung_klimawandel_bf.pdf, 2001.

[FOR 12] FOR PORTFOLIO, Forever open road roadmap, "Climate change resilient transport", available at http://www.foreveropenroad.eu/index.php?m=5, 2012.

[FOR 07] FOR ROADMAP, Forever open road roadmap "Climate change resilient transport", *Climate Change Resilient Transport*, available at http://www.foreveropenroad.eu/index.php?m=6&id_directory=7399.

[ICK 07] ICKERT L., MATTHES U., ROMMERSKIRCHEN S. *et al.*, "Abschätzung der langfristigen Entwicklung des Gütherverkehrs in Deutschland bis 2050", ProgTrans AG, Basel, 2007.

[IPC 07] IPCC, Climate Change, The Physical Science Basis, Contribution of working group *i* to the fourth assessment report of the Intergovernmental Panel on Climate Change, in SOLOMON S., QIN D., MANNING M. *et al.* (eds.), Cambridge University Press, Cambridge, United Kingdom, p. 996, 2007.

[KUM 12] KUMERICS C., KRAUTER E., FEUERBACH J. *et al.*, Abschätzung der Risiken von Hang- und Böschungsrutschungen durch die Zunahme von Extremwetterereignissen, *Berichte der Bundesanstalt für Straßenwesen*, Heft S 75, Bergisch Gladbach, 2012.

[NOV 12] NOVÁK B., ENSLE A., Auswirkungen des Klimawandels auf bestehende Spannbetonbrückenbauwerke, Final report (in German), Bundesanstalt für Straßenwesen, 2012.

[SCH 11] SCHÖNTHALER K., ANDRIAN-WERBURG S., NICKEL D., Entwicklung eines Indikatorensystems für die Deutsche Anpassungsstrategie an den Klimawandel (DAS), Umweltbundesamt, Dessau-Roßlau, available at http://www.uba.de/uba-info-medien/4230.html, 2011.

[TEG 11] TEGETHOF U., BÜRGER M., HARTZ B. *et al.*, Anpassung der Straßenverkehrsinfrastruktur an den Klimawandel (AdSVIS), report (in German), Bundesanstalt für Straßenwesen (BASt), Bergisch Gladbach, 2011.

42

The Impacts of Climate Change on Pavement Maintenance in Queensland, Australia

The research aims to advance the understanding of the impacts of climate change on flexible pavement maintenance in Queensland, Australia. To meet the objective, a case study was carried out to identify the current and projected future climate of a Local Government Authority's road network in Queensland for the next 50 years. Modeling of the road roughness deterioration of flexible pavements was performed using HDM-III road deterioration model under the current and future climate conditions. Using the predicted roughness deterioration data, an economic analysis of the road network pavement maintenance was performed in order to achieve a target pavement condition objective for both the climate scenarios. It was concluded that climate change does play a significant role in increasing pavement deterioration for the local roads in Queensland, Australia. The predominant components contributing to the pavement deterioration are the environmental and climate index. The study has successfully established the relationship between the pavement maintenance cost and the Thornthwaite moisture index. The study concluded that the maintenance cost is projected to increase by nearly 30% under the future climate change condition.

42.1. Introduction

Of late, the topic on climate change has been on the central stage in both national and international forums. Climate change has been recognized globally as an issue of utmost concern and the threat of climate change poses a problem to all nations in the world [IPC 07]. The release of The Garnaut Review 2011 on Australia in the

---

Chapter written by Gary CHAI, Rudi VAN STADEN, Hong GUAN, Greg KELLY and Sanaul CHOWDHURY.

Global Response to Climate Change [GAR 11] serves as a timely reminder that the Australia continent is warming and that this is causing more droughts, water shortages and extreme weather. The report also expressed concerns that Australia is particularly vulnerable to climate change.

Climate change is expected to have severe economic, environmental and social impacts. Particularly vulnerable to these impacts are urban areas, whether inland or coastal, urban areas contain a hub of economic activity and density of human settlement both depending on a reliable and ubiquitous transportation networks. The potential risks and impacts of climate change on these transportation networks pose many challenges to regional development of a country.

Because of its size and location, Australia's climate can be described as diverse and covers a spectrum of climate classifications ranging from pure tropical over the northern quarter to alpine over the highlands of southern New South Wales, Victoria and Tasmania. The country is hot and dry, and as a result the continent is particularly vulnerable to increases in temperature due to climate change. Due to the geography, the country experiences extreme variable climate events ranging from droughts to floods. The climate in Queensland ranges from subtropical in the south to a tropical and equatorial rainforest climate in the north.

The road network in Queensland is about 174,000 km in length. Stewardship of this network lies with two organizations, the Queensland Department of Main Roads and Local Governments. Local Government roads make up approximately 147,000 km of the total road network. The Local Government controlled network is estimated to have a replacement value of more than $10 billion [TAL 03]. This road network is an important physical asset for the state and local governments.

Several countries around the world have examined the expected impacts of climate change on their road infrastructure without modeling deterioration. Galbraith *et al.* [GAL 05] in Scotland, Humphrey [HUM 08] in the United States, Arkell and Darch [ARK 06] in the UK and Kinsella and McGuire [KIN 05] in New Zealand all agreed that projected changes in future climate present new problems for road infrastructure management and design. The general consensus of climate change impacts on road networks are summarized in Table 42.1.

The objective of this chapter is to study the likely impacts of climate change on the pavement maintenance for road network of a Local Government Authority (LGA) in Queensland for the next 50 years based on the predicted climate data. The scopes of the research are as follows:

– evaluating the current and projected future climate for the study area in terms of the climatic indicators;

– modeling of the pavement deterioration using HDM-III road deterioration model for current and future climate conditions;

– perform an economic analysis for the road network pavement maintenance in order to achieve a target pavement condition objective for both the current and climate change scenarios.

Projected climate change	Impact on road network
Rising sea levels	– Coastal erosion – Increase soil salinity levels
Increase surface temperature/solar radiation	– Increased aging and breakdown of asphalt surface – Decrease in material property stiffness – Increased surface deformations
Increased total rainfall and rainfall intensity	– Erosion and soil instability – Overloading of drainage systems causing street flooding – Increased water on road surface
Decrease total rainfall	– Decrease in moisture-related distresses
Increase in Thornthwaite moisture index (TMI, wetter conditions)	– Increased overall deterioration
Decrease in TMI (drier conditions)	– Decreased overall deterioration

Table 42.1. *Climate changes and impacts on road network [KIN 05, HOP 07, DEP 09]*

42.2. Climate change trends in Australia

42.2.1. *Temperature*

Since 1960, the mean temperature in Australia has increased by about 0.7°C [CSI 10]. Also since 1960, some areas in Australia have experienced warming of 0.4°C per decade as shown in Figure 42.1(a), resulting in total warming of 1.5–2°C in the last 50 years. Annual average temperatures in Australia are expected to increase parallel to rises in global average temperature. Temperature increases for inland regions are expected to be greater than coastal regions [SUP 07]. By 2030, annual average temperatures in Australia are projected to increase by 1.0°C relative to 1990 measurements. The range of warming allowing for uncertainty is between 0.6 and 1.5°C. Future projections are more dependent on different emission scenarios as illustrated in Figure 42.1(b). By 2050, annual warming over Australia ranges around 0.8–1.8°C for B1 scenario and 1.5–2.8°C for A1F1 scenario. By

2070, warming ranges from 1.0 to 2.5°C for B1 scenario and 2.2 to 5.0°C for A1F1 scenario [CSI 07].

42.2.2. Precipitation

Precipitation is the quantity of water falling to earth in solid or liquid form (snow, hail and rain) and is not directly influenced by rising greenhouse gases. Increases in temperature and humidity affect the wind circulation and water vapor in the atmosphere, which can change rainfall patterns. These changes are more complex than temperature changes and therefore more uncertainty is taken into account for estimations [SUP 07]. Over the last 50 years from 1960 to 2009, the total rainfall in Australia has been relatively stable; however, the geographical distribution has changed significantly [CSIR 10]. Rainfall has increased in many parts of northern and central Australia, and decreased in south-east and south west Australia as shown in Figure 42.2(a). For 2030, a median decrease between 2 and 5% of annual precipitation with little change in far north Australia is estimated. Differences in emission scenarios cause projected precipitation changes to vary for 2050 and 2070 (Figure 42.2(b)). Overall, it is estimated that most of Australia will decrease in average annual precipitation with expected increases in far northern parts of Australia. Extreme rainfall events are expected to increase in number and intensity [CSI 07, AUS 04].

42.3. HDM-III road deterioration model

HDM-III road deterioration model [WAT 87] is used in this study to model the effects of climate change on flexible pavement deterioration. The projected road conditions generated by HDM-III model form the basis for programing the future pavement maintenance requirements under the climate change scenario. The HDM-III predictive models have been applied in both developed and developing countries having markedly different technology, climatic and economic environments. HDM-III establishes the:

> "causality of events: a pavement starts to crack and to ravel (in a random fashion, after a few years of service); the cracking then increases in extent and intensity; this leads to potholing and other surface disfigurement which, together with rutting, leads to increased roughness – the principal parameter affecting vehicle operating costs" [PAT 87].

HDM-III predicts roughness progression as the sum of three components:

– structural deformation, which is related to roughness, equivalent standard axle load (ESAL) and structural number;

– surface condition, which is related to changes in cracking, potholing and rut depth variation;

– an age–environment-related roughness term.

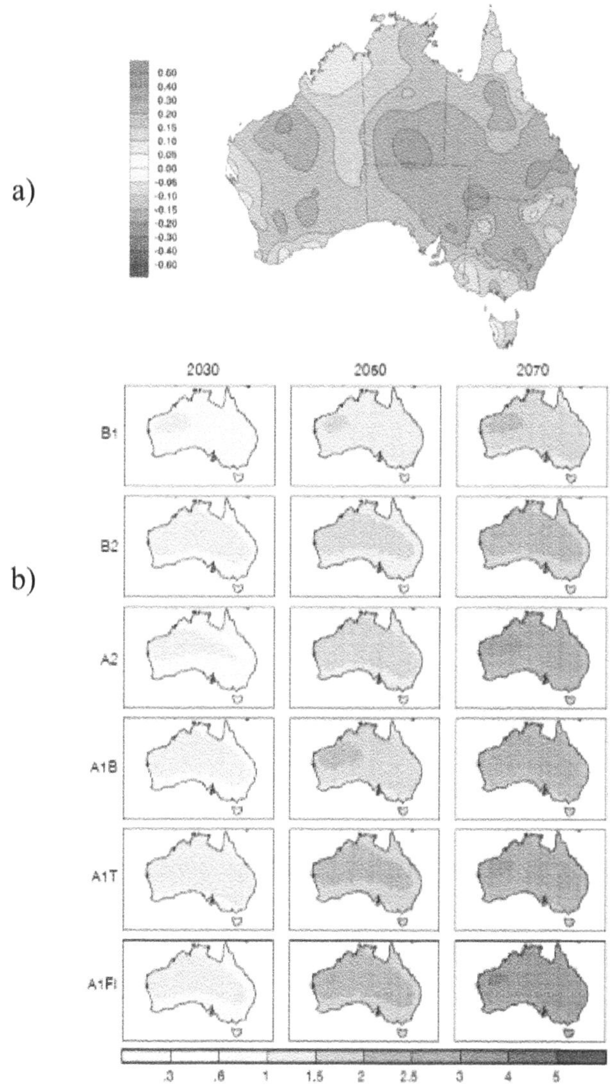

Figure 42.1. *a) Trend in mean temperature 1960–2009 (°C/decade) [CSI 10]; b) Projected annual mean temperatures based on emission scenario for 2030, 2050 and 2070 [CSI 07]. For a color version of this figure, see www.iste.co.uk/jacob/materials.zip*

a)

b)

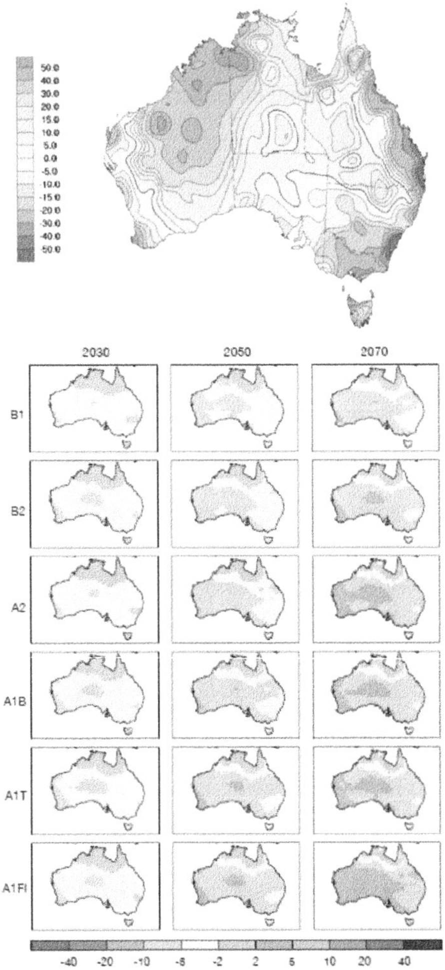

Figure 42.2. *a) Trend in annual rainfall 1960–2009 (mm/decade) [CSI 10]; b) Percentage change in annual precipitation based on emission scenario for 2030, 2050 and 2070 [CSI 07]. For a color version of this figure, see www.iste.co.uk/jacob/materials.zip*

The HDM-III roughness model is expressed using the following relationship [WAT 87]:

$$\Delta\ IRI = K_{gp}[134e^{mt}(SNCK+1)^{-5}YE4 + 0.114\ \Delta\ RDS + 0.0066\ \Delta\ CRX + 0.42\ \Delta\ APOT] + K_{ge}[0.023IRI] \quad [42.1]$$

where ΔIRI is the predicted international roughness index (IRI units), K_{gp} is the user-specified deterioration factor for roughness progression (default value = 1), K_{ge} is the user-specified deterioration factor for environment-related annual fractional increase in roughness (default value = 1), SNCK is the modified structural number for the pavement, reduced for the effect of cracking in asphalt layers, YE4 is the annual axle loading (million ESA per lane), ΔRDS is the annual increment in standard deviation of rut depth (mm) under a 1.2 m straight edge, ΔCRX is the annual increment in indexed cracking (per cent area), ΔAPOT is the annual increment in potholing (per cent area), m is the environmental coefficient, t is the number of years since last major treatment and IRI is the roughness at the start of the year (m/km).

The projected road conditions generated by HDM-III model form the basis for programing the future pavement maintenance requirements under the current and climate change scenario.

42.4. Thornthwaite moisture index

In equation [42.1], the environmental coefficient (m) and the calibration factor for the environmental component of the roughness (K_{ge}) are linked to the Thornthwaite moisture index (TMI). TMI is a climatic characteristic defined by the ratio of the difference between precipitation and evapotranspiration to the potential evapotranspiration. The TMI can be calculated by the humidity index (Ih) minus the aridity index (Ia):

$$\text{TMI} = \text{Ih} - \text{Ia} \qquad [42.2]$$

Expressed as a percentage, the Ih is defined as

$$\text{Ih} = [100\times \text{S}]\ \text{PE} \qquad [42.3]$$

and the Ia is defined as

$$Ia = \underline{[100 \times D]}\ \text{PE} \qquad [42.4]$$

where S is water surplus, D is water deficiency and PE is the potential evaporation.

Thornthwaite [THO 48] found that in an overall moisture index, the Ih has more weight than the Ia; the latter having only six-tenths the value of the former. Annual TMI values are calculated by summing up the monthly TMI values based on the mean monthly temperatures, precipitation and actual evapotranspiration. TMI can be calculated by modifying Thornthwaite's formula [THO 48]:

$$\text{TMI} = [100\times S]/\text{PET} - [60 \times D]/\text{PET} \qquad [42.5]$$

where surplus (S) = precipitation (P) – actual evapotranspiration (AE); deficit (D) = potential evapotranspiration (PET) – actual evapotranspiration (AE).

Modeling of road deterioration was achieved by adjusting the most sensitive deterioration factor, K_{ge}. Note that this factor is the most heavily weighted environmentally related factor. It is observed that a linear relationship exists between the linear rate of roughness deterioration (LRRD) in IRI/year and K_{ge} factor. The relationship is shown as follows (see Figure 42.3):

$$LRRD = 0.1063K_{ge} - 0.0126 \quad [42.6]$$

where LRRD = the linear rate of roughness deterioration in the unit of IRI per year (IRI/year) and K_{ge} is the calibration factor for the environmental coefficient. The expression can be further expanded to link the LRRD with TMI using the following coefficient (m)–TMI relationship:

$$m = 0.0229 + 0.000158 * TMI \quad [42.7]$$

$$K_{ge} = \frac{(m * k_m)}{0.023} \quad [42.8]$$

where m = the environmental coefficient and km = 1. Using equations [42.6], [42.7] and [42.8], the K_{ge}–TMI and LRRD–TMI relationships can be established. The relationships are expressed in equation [42.9], respectively.

$$K_{ge} = \frac{(0.0229 + 0.000158 * TMI)}{0.023}$$

$$K_{ge} = 0.99565 + 0.006869 * TMI \quad [42.9]$$

Substituting equation [42.9] into equation [42.6], the LRRD–TMI relationship can be expressed as:

$$LRRD = 0.1063(0.99565 + 0.006869 * TMI) - 0.0126$$

$$LRRD = 0.0932 + 0.000730 * TMI \quad [42.10]$$

The LRRD–TMI relationship is presented graphically in Figure 42.3. The correlation is considered good with the R^2 value of 0.98. The modeling results show that a direct correlation exists between the rate of roughness deterioration and the moisture index, and the relationship is useful for predicting the impact of TMI on road deterioration. The model shows that an increase of 50% in TMI would result in an increase in the rate of deterioration by the same percentage of increase.

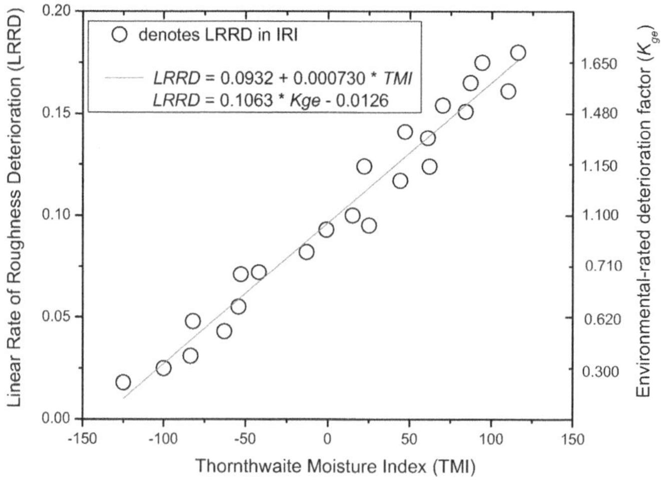

Figure 42.3. *Linear rate of roughness deterioration versus Thornthwaite moisture index*

42.5. Road environment and climate data of the study area

A road network of LGA in Queensland has been selected for the case study. The LGA has a road length of 1,100 km and about 57% of the road network consists of residential access with traffic count less than 1,000 vehicles per day. Approximately 5% of the road network is classified as arterial and subarterial roads with traffic volume greater than 5,000 vehicles per day. Majority of the roads within the LGA (about 95%) would have average daily traffic of less than 5,000 with either thin layer of asphalt and sealed coat. The pavement condition index (PCI) of the road network is 7.5 in a scale from 0 with the worst road condition to 10 being the best condition. PCI of 7.5 is classified in the category of good pavement condition.

Current and future climate data will be defined using the Austroads Climate Tool [AUS 10]. TMI, temperature and rainfall data from global position satellite locations in the vicinity of the road network are used to define climate data. Averages of historical data from 1960 to 2010 are used to define current climate. For future climate trends over 40 years up to year 2050, the Climate Tool uses lower, average and upper severity scenarios based on future emissions. For this study, upper severity is used to examine the worst case scenario that affects the pavement deterioration.

Based on the Australian trends, the future climate in the study area in Queensland is expected to become hotter and wetter (see Table 42.2). The region will experience an increase in mean annual temperatures and increase in mean annual rainfall. The TMI for year 2030 is predicted to be 105.0 and the change in the TMI for the next 30 years is +33.9. The weather is predicted to be wetter with the mean annual rainfall increases to 1,906.4 mm. The region will also experience an increase in mean annual temperature of about +0.8°C over the next 20 years.

Study area in a Local Government in Queensland		
HDM-III input variables	Current climate (1960–2010)	Future climate (2030–2060)
Thornthwaite moisture index	73.61	105.0
Duration of dry season in months	3	3
Mean monthly precipitation (MMP) (mm)	1,780.7	1,906.4
Average monthly temperature range (°C)	28.9	29.7
Number of days greater than 32°C	4	34
Freeze index (not applicable for SEQ)	0	0
Percentage of time driven on water-covered roads	20	20

Table 42.2. *HDM-III climate input variables*

42.6. Economic impacts on pavement maintenance

The results of the pavement modeling for the road network in the study area are presented in Figure 42.4. The road section used in the modeling has an average daily traffic volume (AADT) of less than 1,000 and the ESAL of less than 100,000 per year. The unit used for the roughness deterioration is the IRI [SAY 98]. Figure 42.3 shows the trends of pavement deterioration under the current climate with traffic increases up to year 2035 and future climate change with traffic projected to year 2060. The results of these studies indicate that for higher future, TMI will result in higher pavement deterioration rates. Under current condition (TMI = 66.1), the rate of pavement deterioration is modeled to be 0.115 IRI per year. For a future climate with TMI projected to be 105.0, the deterioration rate is estimated to be 0.170 IRI per year.

The pavement modeling analysis is repeated for the site using a growth rate set to zero percent instead of 4%. This was to ascertain the effect of the traffic growth on the deterioration. The results indicate that the role of traffic increases on the pavement deterioration for the category of road within the LGA is found to be not significant. The predominant factors that contributed to the future pavement deterioration are essentially the environmental and climate factors.

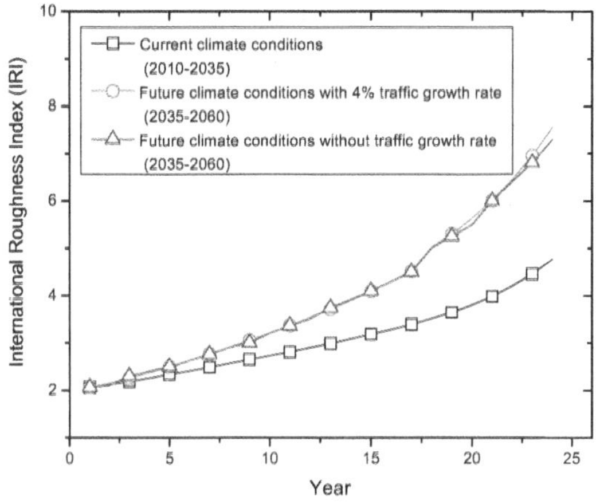

Figure 42.4. *Pavement deterioration at the study area in Queensland*

An economic analysis of the pavement maintenance cost (PMC) for the LGA road network was carried out using the LGA pavement management system (LGA PMS) with HDM-III road deterioration model. The analysis was performed in accordance with the two strategic scenarios with an attempt to minimize investment costs constrained by a target average road network condition (see Table 42.3). The main objective of the analysis was to establish an optimum maintenance cost with a target pavement condition objective. The analysis generated forecast on the annual PMCs in order for the LGA to maintain the current network PCI of 7.5 in a sustainable manner. The strategic analysis was repeated by varying the TMI factor representing the change in the climate conditions in the study area over the next 50 years.

Strategic scenarios	Local government authority	Road network length (km)	Description
1	LGA in Queensland	1,005.0	Minimizing investment costs constrained by a target pavement condition index (PCI) of 7.50. Using current climate condition with TMI of 73.61.
2	LGA in Queensland	1,005.0	Minimizing investment costs constrained by a target PCI of 7.50. Using future climate change condition with TMI of 70.
3–13	LGA in Queensland	1,005.0	Minimizing investment costs constrained by a target PCI of 7.50. Using future climate change condition with TMI of 75, 80, 85, 90, 96.5, 100, 105, 110, 115, 120 and 125.

Table 42.3. *Road management strategic scenarios adopted in the economic analysis*

In the economic analysis, the LGA PMS utilized the target PCI methodology to determine funds required to maintain the network in the present PCI. Scenario 1 maintained the present PCI throughout the analysis period of 25 years until 2035 under the current climate condition (TMI = 73.61). Scenario 8 considers the future climate condition with TMI of 100.0 at year 2060. The required sum of maintenance funds generated by Scenario 1 over the analysis period totalled \$217.75 million. The analysis shows that an average of \$8.71 million of maintenance cost is required per annum to maintain the current PCI under the current climate condition. The annual pavement maintenance expenditure is projected to increase to \$11.31 million per annum when future climate condition (with TMI = 100.0) was adopted as the input parameters in the modeling (see Figure 42.5). The required sum of maintenance funds generated by Scenario 8 over the analysis period is estimated to be \$282.75 million. This represents a change of nearly 30% in the future maintenance cost. A relationship has been established to link the PMC to TMI. The PMC–TMI relationship can be expressed as follows and the equation will be known as TMI cost model:

$$PMC = 4.52 \exp(0.0093 * TMI) \quad [42.11]$$

where PMC = pavement maintenance cost (in Australian dollar).

The quantity of pavement in the road network within the LGA area is 8,543,500 square meters and the PMC per square metre of the pavement is calculated to be \$1.02 and \$1.32 for the current and future climate conditions, respectively. A summary of the PMCs is shown in Table 42.4.

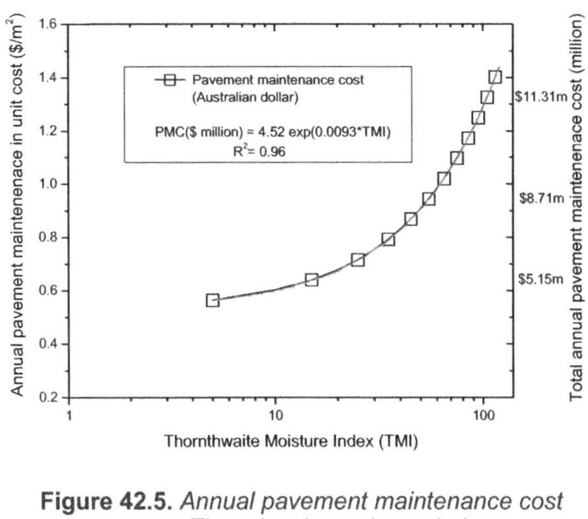

Figure 42.5. *Annual pavement maintenance cost versus Thornthwaite moisture index*

	2014 Base climate	2060 Climate	Change
Annual maintenance cost ($ in million)*	8.71	11.31	+29.8%
Annual maintenance cost ($/m^2)*	1.02	1.32	

*Maintenance cost in Australian dollar.

Table 42.4. *Annual pavement maintenance costs*

42.7. Conclusions

The climate data gathered from the Austroads Climate Tool for the study area indicate increases in mean annual temperatures and increase in mean annual precipitation for the next 50 years. Climate factor as represented by TMI indicates an increase in moisture conditions resulting in wetter, more rainfall conditions for the area in Queensland.

It can be concluded that climate change does play a significant role in increasing rate of pavement deterioration. The predominant components contributing to the future pavement deterioration are the environmental and climate factors. Under current condition (TMI = 73.61), the deterioration rate is modeled to be 0.115 IRI per year. For future climate with TMI projected to be 105.0, the deterioration rate is estimated to be 0.170 IRI per year. The results indicate that the role of traffic increases on the pavement deterioration for the category of road within the LGA is found to be not significant. The pavement deterioration for the category of road (with AADT and ESAL/year of 1,000 and 100,000, respectively) considered in the study is driven largely by the increase in TMI factor and the forecast wetter climate in the LGA area. The study shows that TMI can be useful for assessing future maintenance needs for the pavements surfaced with a fairly thin layer of asphalt or surface dressing and under the temperature and rainfall regimes encountered in the area.

The economic analysis shows that an average of $8.71 million of maintenance cost is required per annum to maintain the existing PCI under the current climate condition. The annual pavement maintenance expenditure is projected to increase to $11.31 million when future climate condition was adopted as the input parameters in the modeling. This represents an increase of nearly 30% in the future maintenance cost.

Climate change will present challenges to road engineers because of the large number of implications for the design, construction and maintenance of road pavements. Highways and roads are designed based on moisture and temperature patterns reflecting the history of the local climate. With projected climate changes to accelerate over the next several decades, pavements will be subjected to very different climatic conditions over the design life than was originally expected. With the level of information, LGAs are in a far more informed position to forecast future road funding requirements and can better serve the community by maintaining an adequate level of overall road conditions. Additionally, the study provides LGAs with a new approach using TMI climate index for predicting the cost consequences in case of resilience is not being built-in the road infrastructure.

42.8. Acknowledgments

The authors would like to thank the South East Queensland Climate Adaptation Research Initiative (SEQ-CARI) for providing funding for the research. The authors would also like to thank Queensland Department of Transport and Main Roads (Road Asset Management Branch, South Coast & Brisbane Metropolitan Regions), Gold Coast, Redland and Logan City Councils, and the Southern Regional Roads Group for the financial support. Their funding contribution is greatly appreciated.

42.9. Bibliography

[ARK 06] ARKELL B.P., DARCH G.J.C., "Impact of climate change on London's transport network", *Proceedings of the Institution of Civil Engineers, Municipal Engineering*, vol. 159, no. ME4, pp. 231–237, 2006.

[AUS 04] AUSTROADS, Impact of climate change on road infrastructure, Publication AP-R243, Austroads Ltd., Sydney, 2004.

[AUS 10] AUSTROADS, "Impact of climate change on road performance", *Updating Climate Information for Australia*, Austroads Ltd., Sydney, 2010.

[CSI 07] CSIRO & BOM, Climate change in Australia: technical report 2007, Commonwealth Scientific and Industrial Research Organisation & Australian Bureau of Meteorology, CSIRO, Melbourne, 2007.

[CSI 10] CSIRO & BOM, State of the climate – snapshot, Commonwealth Scientific and Industrial Research Organisation & Australian Bureau of Meteorology, CSIRO, Melbourne, 2010.

[DEP 09] DEPARTMENT OF CLIMATE CHANGE, Climate change risks to Australia's coast: a first pass national assessment, Commonwealth of Australia, 2009.

[GAL 05] GALBRAITH R.M., PRICE D.J., *Scottish Road Network Climate Change Study*, The Scottish Government, Glasgow, 2005.

[GAR 11] GARNAUT R., *The Garnaut Review 2011– Australia in the Global Response to Climate Change*, Cambridge University Press, Port Melbourne, Victoria, Australia, 2011.

[HOP 07] HOPLER P., LUCY S., NOLAN M. *et al.*, Infrastructure and climate change risk assessment for Victoria, Report to the Victorian Government, CSIRO, 2007.

[HUM 08] HUMPHREY N.P., TRB special report: potential impacts of climate change on US transportation, committee on climate change and U.S. transportation, Transportation Research Board, Division on Earth and Life Studies, National Academy of Science, 2008.

[IPC 07] IPCC, Climate Change, Synthesis Report. Contribution of Working Groups I, II and III to the Fourth Assessment Report of the Intergovernmental Panel on Climate Change, PACHAURI R.K., REISINGER A. (eds.), IPCC, Geneva, Switzerland, p. 104, 2007.

[KIN 05] KINSELLA Y., MCGUIRE F., *Climate Change Uncertainty and the State Highway Network: A Moving Target*, Transit New Zealand, 2005.

[MIL 09] MILLS B.N., TIGHE S,L., ANDREY J. *et al.*, "Climate change implications for flexible pavement design and performance in southern Canada", *Journal of Transportation Engineering*, vol. 135, no. 10, pp. 773–782, 2009.

[SAY 98] SAYERS M.W., KARAMIHAS S.M., *The Little Book of Profiling – Basic Information about Measuring and Interpreting Road Profiles*, The Regent of the University of Michigan, Ann Arbor, Michigan, USA, 1998.

[SUP 07] SUPPIAH R., HENNESSY K.J., WHETTON P.H. *et al.*, "Australian climate change projections derived from simulations preformed for the IPPC 4th Assessment Report", *CSIRO Marine and Atmospheric Research*, Aust. Met. Mag, vol. 56, pp. 131–152, 2007.

[TAL 03] TALBOT S., PELEVIN A., "The Queensland asset management partnership", *Proceedings of 21st ARRB & 11th REAAA Conference*, Australian Road Research Board (ARRB), Cairns, Australia, May 18–23, 2003.

[THO 48] THORNTHWAITE C.W., "An approach toward a rational classification of climate", *The Geographical Review*, vol. 38, pp. 55–94, 1948.

[WAT 87] WATANATADA T., PATERSON W.D.O., BHANDARI A., *The Highway Design and Maintenance Standards Model; Volume 1. Description of the HDM-III. World Bank Publication*, John Hopkins University Press, Baltimore/London, 1987.

[PAT 87] PATERSON W.D.O., *The Highway Design and Maintenance Standards Series; Road Deterioration and Maintenance Effects. World Bank Publication*, The John Hopkins University Press, Baltimore/London, 1987.

43

Design Guideline for a Climate Projection Data Base and Specific Climate Indices for Roads: CliPDaR

The mission of CliPDaR is to issue a guideline setting, a standard regarding the handling of climatological data and methods that shall serve as a basis for pan-European traffic infrastructure risk assessments. This includes a stepwise description of the ensemble approach starting from socioeconomic scenarios over global climate models (GCMs) and selected downscaling methods to generate regional scale climate change projections, which can be used to drive impact models. Within this study, we present this approach by calculating three climate indices that are associated with damage to road surface, supporting structure and drainage systems. These indices are calculated for 20 spots of the Trans-European Transport Network and two future periods (2021–2050 and 2071–2100) and compared to conditions of a time slice "1961–1990" from the so-called "control run".

43.1. Introduction

Traffic infrastructure is of the upmost importance to economy as well as to people. The supply of daily goods or the accessibility of hospitals, for instance, heavily relies on the trafficability of roads all year long. Today, about 70% of the total freight is carried across roads and this number is expected to significantly increase in the decades to come. The volume of traffic is estimated to grow by 85% from its current value within the next 25 years. Aside from this enormous increase, there are other challenges to future road networks that have to be considered such as climate change, demographic development, or new advances in technology. All these changes will affect road infrastructure elements. Changing needs for

---

Chapter written by Christoph MATULLA, Joachim NAMYSLO, Konrad ANDRE, Barbara CHIMANI and Tobias FUCHS.

maintenance and reinforcement works require far-sighted planning. Rutting of asphalt surfaces or "blow ups" of concrete roads are safety issues. They are related to climate indices (CIs) characterizing heat days coming along with tropical nights, which may become more frequent in the future. Changes in the frequency of freeze-thaw cycles or altering precipitation patterns result in different profiles of risks to road surfaces, slope support or drainage systems.

Bridges, tunnels, supporting structures, culverts, slope protection measures, road surface, drainages, and pump systems are stationary assets belonging to the road network. If an individual asset fails, the whole system is at risk. As such, it is important to know the probability of damages to such infrastructure elements. In this study, we focus on climate-related risks affecting maintenance budgets. In this context, it is relevant to note that strategic decision making on transport issues (planning and designing, construction and reinforcement works and substantial changes in maintenance and budget strategies) refers to periods of some decades, which is the characteristic time scale on which climate change emerges. Infrastructure cycles are on the same time scale as climate change. Hence, climate change should be considered in today's plans for future transport networks.

This may be illustrated by a simple thought experiment: rutting, for instance, is controlled by the weight of crossing vehicles and road surface temperatures. If surface temperature values exceed a certain threshold (e.g. 55°C), the risk of damage increases significantly. In case of an increasing summer temperature, this threshold will be exceeded more often and the economic loss might not be acceptable anymore. To avoid such a situation, it might be necessary to alter the asphalt mix in the design of new roads right now.

Similar developments could be expected with several other CIs (e.g. precipitation events exceeding a certain threshold). As it is less expensive to include possible effects of climate change in the planning than to adapt them later, it is mandatory to consider climate change already in the course of planning.

43.2. Data and methods

43.2.1. *Data*

As CliPDaR is to support decision-making regarding future transport infrastructure, datasets used to describe changes in CIs have to cover past and possible future climatic conditions. Further requirements refer to the spatiotemporal resolution, permitting a more or less detailed view. Here, we make use of eight members of the KLIWAS ensemble [IMB 13], describing past and future conditions up to the end of the 21st Century. KLIWAS-8 is an eight-member ensemble based on regional climate model (RCMs, Table 43.1) projections providing daily values of mean temperature, precipitation sum, relative humidity and sum of global radiation on a 25 km grid. These climate variables are statistically downscaled to a 5kmgrid and biascorrected. The spatial (5 km) resolution of the KLIWAS-8 ensemble is exceptional. This is essential for a profound downscaling of climate (projection) data to selected parts of the road infrastructure, such as bridges or box-cuts by the help of local ("microscale") thermodynamic climate models.

Daily maximum and minimum temperatures are approximated by a parameterization function driven by daily mean temperature, ratio of daily actual over daily maximum global radiation and by a sinusfunction of a daily maximum temperature range from 4°C in the winter period up to 14°C in the summer period, drawn from climate statistics of averaged measured extreme temperature data for a "typical" Mid-European climate station.

KLIWAS is based on the socioeconomic scenario abbreviated as "A1B" [IPC 07] that describes a future world of fast economic growth, a rapid spread of innovative and efficient technology around the globe, a population that increases until the middle of the century and decreases afterward and an energy consumption that is supplied by fossil as well as non-fossil energy sources to equal parts.

Control run/SRES scenario	GCM	RCM	No.
C20/A1B	HadCM3Q0 (HC)	CLM2.4.6 (ETHZ)	1
		HadRM3Q0 (HC)	2
	BCM2 (NERSC)	RCA3 (SMHI)	3
	ECHAM5-r3 (MPI-M)	RegCM3 (ICTP)	4
		HIRHAM5 (DMI)	5
		RACMO2 (KNMI)	6
		REMO5.7 (MPI-M)	7
	ECHAM5-r1 (MPI-M)	CLM2.4.11 (GKSS)	8

Table 43.1. *Overview of climate simulations of (i) the years 1961–2000 for the control run (C20), (ii) projection runs for the years 2001–2100 based on the scenario A1B (special eight-member ensemble (5 km scale) of daily values used in KLIWAS, original data 25 km scaled; after [IMB 13], modified). GCM, global climate model*

43.2.2. *Methods*

43.2.2.1. *Cause–effect relations*

Relationships between CIs and road infrastructure elements are of central importance. An objective way to isolate them is to analyze the functional dependency between time series of damage and climatological variables (e.g. road surface damages and temperature evolution). Another way involves expert knowledge on physical processes and experience. As no time series of damage were available, expert knowledge was gained through workshops and interviews. Figure 43.1 lists infrastructure elements together with CIs potentially harming them. Concerning climate change, this "cause–effect tensor" (CET2) is the very center of the whole task scope ensuring the smooth functioning of transport networks in the future.

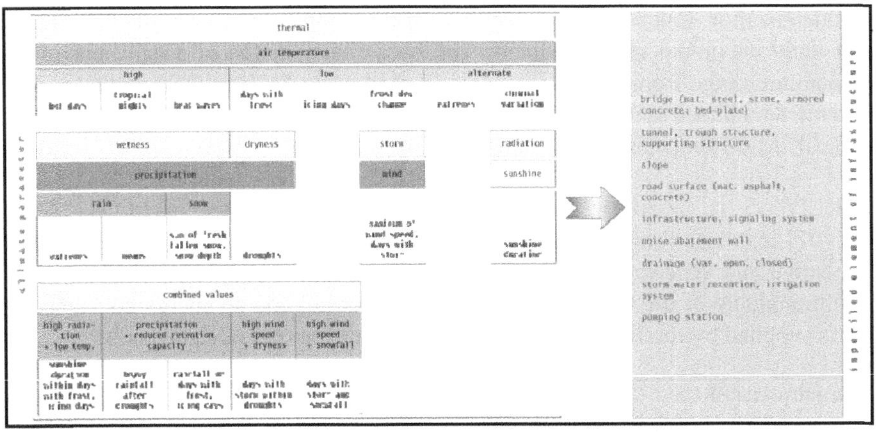

Figure 43.1. *Cause–effect tensor (CET2). Left: climatological elements, right: traffic infrastructure (example) [KOR 15]*

The opportunity to estimate changing risks depends on the capacity of climate change projections. In the following, the generation of local scale climate change projections is elaborated. Therefrom, the future behaviors of the CIs are calculated, which are needed to assess the impact on transport infrastructure. CIs vary in space, time and complexity. Depending on the climatic phenomena, CIs can be valid for regions extending from hectares to thousands square kilometers and can be made up of one or more parameters averaged over different periods of time.

43.2.2.2. *The "downscaling cascade" and the appendant uncertainty*

Since it has become more and more evident that mankind is partly responsible for the observed climate change (see the succession of the IPCC assessment reports), the demand for climate change projections increased continuously. In order to assess the impact of climate change taking place on the regional scale (e.g. transport networks), it is indispensable to work with climate change projections on small scales. The generation of regional scale climate change projections and the subsequent impact assessment involve several methodological steps and datasets.

Figure 43.2 shows the whole process, which is part of CliPDaR, and will be elaborated step by step for three CIs (responsible for rutting, flooding and damage to road surfaces caused by frost-thaw cycles) in the following.

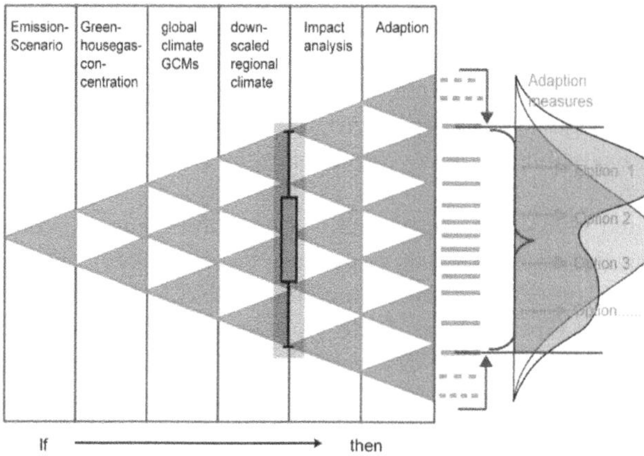

Figure 43.2. *Starting from a particular emission scenario (explained in the text), the uncertainty grows with every step that is required to derive different adaptation measures mitigating the impact of climate change (schematic diagram) after [VIN 02]. The two curves on the right are representative of all possible frequency distributions of climate change impacts*

The process starts with the selection of a so-called socioeconomic scenario (indicated by the leftmost column in Figure 43.2). Ever since it was demonstrated that mankind affects climate by altering the chemical composition of the atmosphere, questions about the consequences of contrasting human behaviors were raised. Possible future pathways of mankind are presented as socioeconomic scenarios that are based amongst others on assumptions about the political, demographic, technological development, changes in land use patterns and how the energy demand is met through the 21st Century [IPC 07].

Socioeconomic scenarios are translated into emission scenarios, telling the time-dependent release of greenhouse gases into the atmosphere, coming along with a particular pathway of mankind. These emissions are translated via gas-cycle models into concentrations (second column from the left), forcing the climate system via the radiation transfer in the atmosphere. This step introduces uncertainties coming from the application of models, which are an approximation to the physical processes. Different models give rise to somewhat different greenhouse gas concentrations.

The next step is to drive global climate models (GCMs) with the temporal development of greenhouse gas concentrations in the atmosphere but constant external radiative forcings. GCMs are three-dimensional numerical approximations of the Earth consisting of climate components (e.g. atmosphere, ocean and cryosphere). They simulate processes within and in between the components, which take place on very different spatiotemporal scales. Different GCMs, designed at various climate research centers around the globe, produce a diversity of climate change projections. This means that the same socioeconomic scenario results in somewhat different global scale climate change projections. This is indicated by a further increasein the amount of uncertainty in Figure 43.2. A climate change projection is one possible future development of the climate.

The simulated global climate projections pose the boundary conditions for the RCMruns. Hence, the next step is to cascade the GCM projections down from a continental scale (200–350 km) to the regional scale (25–50 km). This can be done by a strategy called downscaling. There are essentially two approaches: statistical and dynamical downscaling. The downscaling step from the continental scale to the regional scale introduces yet more uncertainties (see Figure 43.2). Different downscaling techniques have their pros and cons, depending on which climate variable shall be analyzed.

Up to this point in the assessment of possible impacts that mankind may exert on regional scale ecosystems or economic structures (e.g. transport networks), quite some uncertainty has accumulated. This should not be seen as a drawback. In fact, the span represents a variety of possible local scale climate change reactions that may come along with a specific development of mankind. This means that all conclusions on the "adaptation level" depend on "if–then relations", achieved by applying reasonable models. Because the "emission scenario level" reflects only a more or less reasonable statement, results of following levels can only be checked for their plausibility from a climatological or geophysical point of view. That is why so-called "no regret or low regret actions" are frequently proposed in national strategies for adaptation to climate change. The following steps in Figure 43.2 describe (i) how a particular regional climate change projection affects systems (e.g. the surface temperature of road surfaces) and (ii) what measures may be set to manage the impact (e.g. research for new materials or other adaptation actions). These steps are reached by impact models introducing further uncertainty.

43.3. The ensemble approach

With the accelerating advancement in computer technology (increasing velocity of the computational infrastructure and the storage facilities), it became possible to consider several local scale climate change projections based on a number of GCM runs when investigating future climate states, instead of just using one GCM realization. Such a set of projections is called an "ensemble". An ensemble is described by the statistics (median, variation, spread, etc.) of its members (= projections). There are different kinds of ensembles, depending on the question under consideration:

i) "initial condition ensembles": based on same model and emission scenario but different initial conditions;

ii) "multimodel ensembles": different models but the same scenario (Figure 43.3);

iii) "multimodel multiscenario ensembles": different models and scenarios.

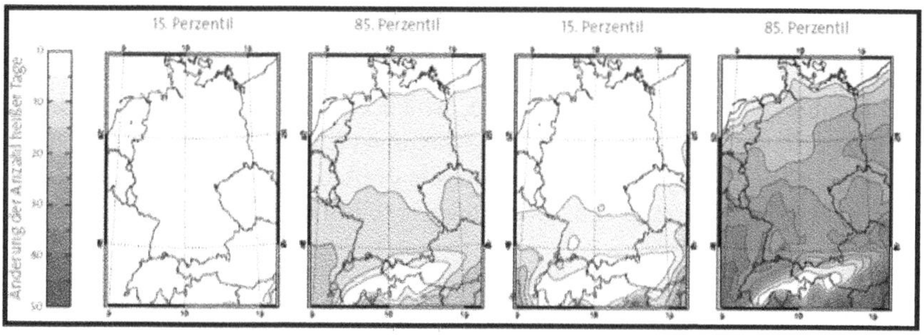

Figure 43.3. *One possibility to illustrate the findings from an ensemble of projections (a multimodel ensemble for A1B). The panels show the increase in the number of hot days per year. The left two panels refer to the 15th and 85th percentiles for the n.f., the right two panels for the f.f. (source: DWD,www.dwd.de/klimaatlas). For a color version of this figure, see www.iste.co.uk/jacob/materials.zip*

Table 43.1 lists the KLIWAS-8 ensemble based on different GCMs and downscaling methods (RCMs). Assuming a meaningful sample size, the median of an ensemble is rather stable against outliers (=climate projections yielding values far from the others). Therefore, the median of an ensemble of projections is rated higher, when approximating the probable future state of the climate system, than a single projection alone. Next to the median, the average and the variation among the ensemble members are of importance (as well as further statistical features of the probability distribution of the ensemble). Experience indicates that different GCMs introduce more variability into a regional scale ensemble than different downscaling models do. Consequently, a multi-GCM ensemble together with several downscaling methods should exhibit enhanced bias values compared to an initial condition ensemble.

Figure 43.3 shows one possible way to display ensemble results (here for the annual number of hot days ($T_{max} \geq 30°C$). The 15th and the 85th percentiles are shown for two future periods (2021–2050 and 2071–2100). Thus, 70% of the projections are in between the panels (the probability to draw a projection from the ensemble that gives values between those shown in the panels is 70%). Figure 43.3 relies on the full KLIWAS ensemble [IMB 13].

43.4. Results and discussion

43.4.1. *General comments*

Within this section, the CIs, explained and motivated above, are presented. They are significant examples taken from the CET2 shown in Figure 43.1. The first CI (Figure 43.4) refers to freeze-thaw cycles, which are responsible for quite a range of damages to transport infrastructure elements including rock fall and cracks in the road surface (leading to consequential damages). The second one refers to precipitation events of and above 30 mm, which affect drainage systems. The third CI describes days made up by heat during daytime (T_{max}>30°C) followed by tropical nights (T_{min}>20°C). This calculation process is drafted in the fifth column of Figure 43.2 and completes the approach to be guided by CliPDaR.

The result values are displayed in a so-called boxplots, giving a comprehensive view on the data. Bottom and top of the box are the first and third quartiles; the blue line inside the box is the average over the valueand the red line with the diamonds on it is the second quartile (i.e. the median). The vertical lines, called whiskers, are 1.5 times the interquartile range. In case there are no results showing deviations that are far from the median, the whiskers end at the maximum deviations. Triangles indicate result values (outliers) being farther away from the median than 1.5 times the interquartile range (25–75%). High interquartile ranges embody more dispersion of the results than small ranges. In the context of the CIs, high ranges indicate rather low levels of consensus among the ensemble realizations, which in turn points to enhanced uncertainty. Low interquartile ranges on the other hand hint to confidence.

Figures 43.4–43.6 show three groups of boxplots. The leftmost group refers to a past period, the middle one to the near future (n.f.) (2021–2050) and the right group to the farther future (f.f.) (2071–2100). Each local scale realization of KLIWAS-8 is depicted by one boxplot just as the mean of the ensemble, which is matched by the filled lightred boxplot.

The CIs are evaluated for 20 "transport spots" in Europe. The calculation of the CIs for the transport spots involves nine points of the KLIWAS-8 grid. So, the transport spots reflect conditions representative for small regions, named after the city exitedinside these regions. The transport spots and the associated cities are listed in Table 43.2.

No.	Name	Lon (°E)	Lat (°N)	Alt (m)	No.	Name	Lon (°E)	Lat (°N)	Alt (m)
1	Munich	11.58	48.15	517	11	Berlin	13.39	52.52	36
2	Stuttgart	9.20	48.80	335	12	Praha	14.44	50.08	327
3	Passau	13.41	48.58	303	13	Linz	14.32	48.30	561
4	Salzburg	13.05	47.81	420	14	Zurich	8.54	47.37	478
5	Innsbruck	11.40	47.27	581	15	Nuremberg	11.08	49.45	398
6	Frankfurt/Main	8.68	50.11	106	16	Dresden	13.74	51.05	140
7	Cologne	6.96	50.94	42	17	Hannover	9.73	52.38	58
8	Dortmund	7.45	51.50	107	18	Leipzig	12.38	51.34	119
9	Amsterdam	4.90	52.37	0	19	Brno	16.62	9.18	313
10	Hamburg	10.00	53.54	15	20	Kiel	10.15	54.38	12

Table 43.2. *Transport spots for which the climate indices are calculated. This calculation involves nine points of the KLIWAS-8 ensemble considered for averaging*

43.4.2. *Distinct zero temperature crossings (freezing-thaw crossings)*

Figure 43.4 shows the number of distinct zero temperature crossings. The label "distinct" means that only days featuring $T_{min} \leq -2°C$ and $T_{max} \geq 2°C$ are considered. This temperature interval should very roughly signify the availability of enough energy to enforce the phase transition from liquid to solid or back. The overall message, which is present in all KLIWAS-8 ensemble members, is that freeze-thawchange (FTC) is projected to decrease in the future. The variance of the mean over all spots decreases from 4.5 to 1.7 between 1961–1990 and 2071–2100. Decreases are somewhat more pronounced in the east and south of the domain.

In 1961–1990, most "distinct" zero temperature crossings are to be found in the east and south of the domain. The result for Innsbruck is different from the results of the neighboring traffic spots, which show more crossings (Table 43.3). A comparison of FTC calculated from measurements for Innsbruck and Linz shows that the low FTC value calculated from the KLIWAS-8 ensemble does not match reality. The too low FTC value in Innsbruck may be caused by the coarse model topography, the location of Innsbruck close to the edge of the model domain or the effect of irradiation, which is not realistically reproduced (perhaps due to the valley character). This example shows the necessity of validating the model output against observations in order to be able to judge the model output (e.g. the generated scenarios) in the right way.

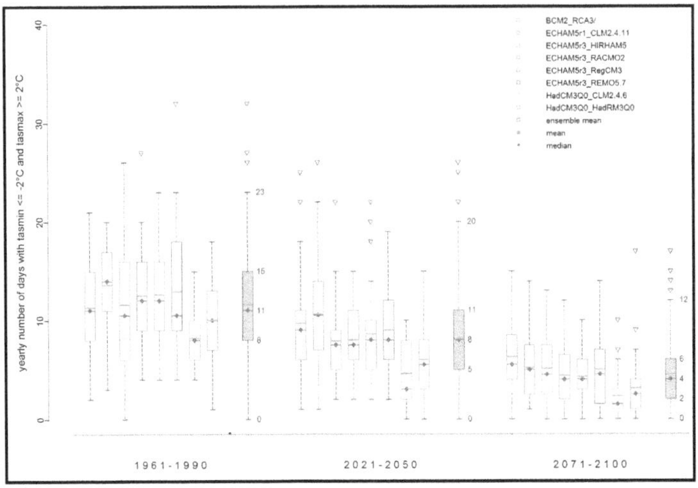

Figure 43.4. *Boxplots of the yearly number of FTC for Praha (a detailed explanation of the plot is given in the "General comments" section). For a color version of this figure, see www.iste.co.uk/jacob/materials.zip*

The differences between the ensemble members, and therefore the uncertainties, vary between the traffic spots and hence through space. No single ensemble member is suspiciously different from the others at all transport spots. In some cases, enhanced differences between the HadCM-driven model runs and the other ensemble members can be seen, as in the case of Praha (Figure 43.4), where the median HadCM FTC value is about 4 days (which is about 50%) below the other models. Single ECHAM runs, however, can differ significantly as well (not shown). The structure of the differences between the ensemble means does not have to be the same in all three considered periods.

43.4.3. *Precipitation totals above 30 mm*

The boxplots showing the number of precipitation events of and above 30 mm are located at low values. This applies to all spots and the whole ensemble indicating a rather robust signal. For the past period, highest values (3–4 days/year) are to be found in Salzburg (see Figure 43.5). The mean over all traffic spots and the whole ensemble is 0.6, which is increasing toward the end of the century to 1 day per year. In case of Salzburg, the increase (i.e. 1 day per year) is more pronounced, yielding 4–5 days per year for the period from 2071 to 2100. The spatial pattern of the increasing values outlines a region at the northern edge of the Alpine chain featuring largest growing rates. This region is prominent for precipitation events triggered by the advection of air masses from the North-West climbing the mountain ridges

("Staulagen"). So, the increase may point to a more often occurrence of such meteorological events. The interquartile ranges do not change with time, pointing to unchanged confidence as measured by the behavior of the projections among each other. No significant changes in the mean and variance of the ensemble members can be found between 1961–1990 and 2071–2100 (Table 43.4).

FTC	Median of ensemble means			FTC	Median of ensemble means		
No.	Past (1961–1990)	n.f.	f.f.	nr.	Past (1961–1990)	n.f.	f.f.
1	10.9	8.0	4.6	11	8.7	5.6	3.1
2	8.5	6.4	2.9	12	10.9	8.0	4.1
3	10.7	7.9	4.1	13	15.0	10.8	6.5
4	9.2	6.5	3.6	14	9.3	7.0	3.2
5	6.7	4.7	1.9	15	11.3	8.0	4.6
6	7.3	5.1	2.5	16	8.9	6.3	3.0
7	6.3	3.5	1.6	17	8.9	5.9	2.7
8	6.7	4.5	2.1	18	9.4	6.0	3.1
9	5.0	3.4	1.4	19	11.9	8.5	4.8
10	8.9	5.8	2.8	20	8.4	5.2	2.1

Table 43.3. *Median for FTC of the KLIWAS ensemble mean for 20 sites and three time period*

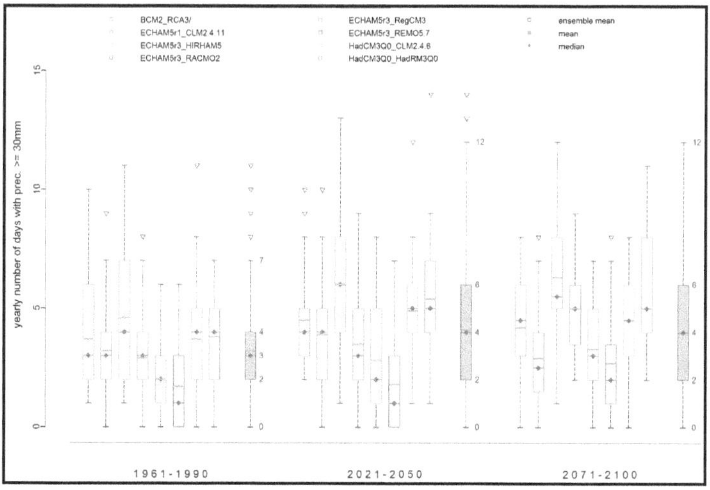

Figure 43.5. *Boxplots for the yearly number heavy precipitation days (precipitation amount ≥30 mm) for Salzburg (a detailed explanation of the plot is given in the "General comments" section). For a color version of this figure, see www.iste.co.uk/jacob/materials.zip*

Parameter	Mean pa./f.f.	Significant	Variance pa./f.f.	Significant
FTC	–6.0	Yes	–3.2	No
RR	0.5	No	0.5	No
TRUT	6.3	Yes	6.3	Yes

Table 43.4. *Differences in mean and variance of the model results averaged over all sites and the appendant significance (on the 0.95-level) of the changes for the three CIs. The time periods 2071–2100 and 1961–1990 have been compared*

43.4.4. *Temperature indices isolating meteorological phenomena causing rutting*

Temperature-related CIs cover a large proportion of CET2 (Figure 43.1) and hence are responsible for a multitude of damages to transport infrastructure. Here, we focus on a CI (henceforth called TRUT) that is characterized by high daily temperatures (≥30°C, hot days) together with T_{min} ≥20°C (tropical nights). Such days bear the potential of harming road surfaces.

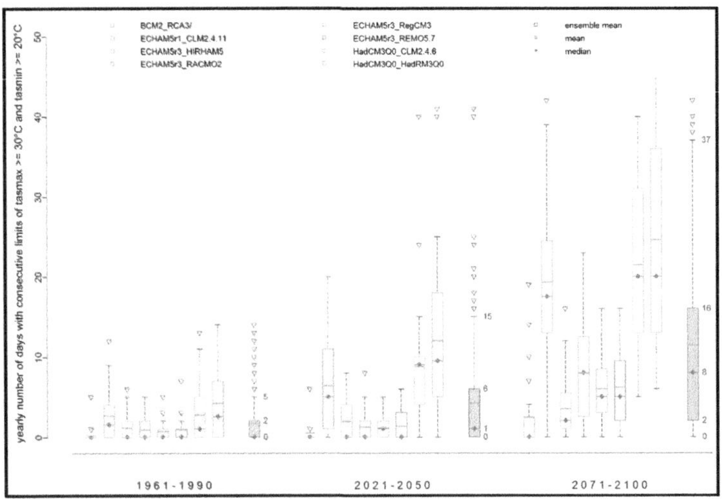

Figure 43.6. *Boxplots for the yearly number of potential "rutting days" (T_{max}≥30°C and T_{min}≥20°C) for Frankfurt am Main (a detailed explanation of the plot is given in the "General comments" section) Low occurrences appear near zero-level. For a color version of this figure, see www.iste.co.uk/jacob/materials.zip*

Presently, such days are rather rare throughout the parts of Europe considered (Figure 43.6). Taking the average over the median at all transport spots of KLIWAS-8, less than one event per year is to be expected. Frankfurt am Main shows the highest frequency (1.6 occurrences/year). This number increases to 7.4 days/year in 2071–2100. This time slice, however, exhibits pronounced differences within KLIWAS-8. This behavior is particularly apparent for the two HadCM runs and the CLM run driven by ECHAM5 (see Table 43.1). While these three projections point to 14 occurrences per year, the other five projections give two days per year. This difference is to be seen at most sites, slightly more pronounced in the south than in the north of the domain and already detectable (albeit distinctly less expressed) in the n.f. (2021–2050). The variance of KLIWAS-8 changes significantly (Table 43.4) between 1961–1990 and 2071–2100, meaning that the uncertainties increase toward the end of the century too.

43.5. Conclusions

Changes in temperature-based CIs are more reliable than those based on precipitation. The evolution of FTCs in the future indicates decreasing damages from appendant events. These, however, may be outbalanced by increases in the TRUT frequency. Which effect will have more impact on the transport infrastructure varies between regions. Even the slight changes in the number of days with heavy precipitation might exert a noticeable effect in some regions and may further depend on changes in the intensity of the precipitation (which was not analyzed in this chapter). It will need expert knowledge of road authorities to decide on the measures to be taken in the different regions for maintenance and reinforcement of road infrastructure.

43.5.1. *Outlook*

The actual work has its focus on the evaluation of the three calculated CIs for the reference time period 1961–1990 with HYRAS, a gridded data set based on observed conditions with a 5 km grid spacing [RAU 13]. Therefore, the climate signals, as differences of the number of days of the future time periods to the reference period, have to be added to the number of days drawn from HYRAS for each selected CI. These results will be compared with measured data from representative stations.

It is still necessary to identify further CIs harming road assets in cooperation with the road administrations, people in charge and constructional engineers. Next to this, KLIWAS and VALUE as well as the German Adaptation Strategy (DAS), the Austrian Adaptation Strategy and the IPCC Recommendations [IPC 07] regarding adaptation measures will be taken into account.

Furthermore, the new outcomes of the 5th report of the IPCC with the next generation of (global and) regional climate projections with higher spatial resolutions in space (up to 10 km) and time (up to hourly values) based on the representative concentration pathway scenarios will be used to enlarge the given ensembles in the followingyears.

43.6. Acknowledgments

The research within CliPDaR is carried out as part of the CEDR Transnational Road research Programme Call 2012. The funding for the research is provided by the national road administrations of the Netherlands, Denmark, Germany and Norway. The authors thank the KLIWAS group of DWD, especially Ulf Riediger for preparing the data of actual daily temperature ranges and Florian Imbery for helpful discussions on data analysis. We would also like to thank Sabrina Plagemann and Florian Imbery for providing the KLIWAS-8-Ensemble (downscaling and biascorrection). We further express our gratitude to Beate Gardeike (HZG), Christine Hagen and Nathalie Nosek (both ZAMG) for their valuable assistance in preparing this manuscript and large parts of the overall CliPDaR dissemination. Furthermore, we are grateful to anonymous reviewers for their helpful comments.

43.7. Bibliography

[IMB 13] IMBERY F., PLAGEMANN S., NAMYSLO J., "Processing and Analysing an Ensemble of Climate Projections for the Joint Research Project KLIWAS", *Advances in Science and Research*, vol. 10, pp. 91–98, 2013.

[KOR 15] KORN M., LEUPOLD A., MAYER S. *et al.*, Risikoanalyse wichtiger Verkehrsachsen des Bundesfernstraßennetzes im Kontext des Klimawandels (RIVA – Risk analysis of key transport corridors of the German motorway network in the context of climate change), Schlussbericht im Auftrag der Bundesanstalt für Straßenwesen, Final Report, 2015.

[PCC 07] PCC, *Climate Change 2007: The Physical Science Basis, Contribution of Working Group I to the Fourth Assessment Report of the Intergovernmental Panel on Climate Change*, Cambridge University Press, Cambridge, UK, 2007.

[RAU 13] RAUTHE M.,STEINER H., RIEDIGER U. *et al.*, "A Central European precipitation climatology – Part I: Generation and validation of a high-resolution gridded daily data set (HYRAS)", *Meteorologische Zeitschrift*, vol. 22, pp. 235–256, 2013.

[VIN 02] VINER D.,"A qualitative assessment of the sources of uncertainty in climate change impacts assessment studies", *Climatic Change: Implications for the Hydrological Cycle and for Water Management*, Advances in Global Change Research, vol. 10, pp. 139–149, 2002.

List of Authors

Scott ACTON
University of Virginia
Charlottesville
USA

Adewole ADESIYUN
Forum of European National Highway Research Laboratories
Brussels
Belgium

Gordon AIREY
Nottingham Transportation Engineering Centre
Department of Civil Engineering
University of Nottingham
UK

Alexander ALISOV
Institut für Straßenwesen
Technische Universität Braunschweig
Germany

Dominique ALLAGNAT
Egis-Géotechnique
Grenoble
France

Konrad ANDRE
Central Institute for Meteorology and Geodynamics (ZAMG)
Vienna
Austria

Pierre ARGOUL
French Institute of Science and Technology for Transport, Development and Networks (IFSTTAR)
Marne-la-Vallée
France

Markus AUERBACH
Bundesanstalt für Straßenwesen
Bergisch Gladbach
Germany

Lidija AVSENIK
National Institute of Chemistry
Ljubljana
Slovenia

J. Stefan BALD
Technische Universitaet Darmstadt
Germany

Fátima BATISTA
Laboratório Nacional de
Engenharia Civil
Portugal

Abdeldjelil BELARBI
University of Houston
Texas
USA

Hans BENDTSEN
Danish Road Directorate (DRD)
Copenhagen
Denmark

Karim BENZARTI
French Institute of Science and
Technology for Transport,
Development and Networks
(IFSTTAR)
Marne-la-Vallée
France

Anneleen BERGIERS
Belgian Road Research Centre
(BRRC)
Brussels
Belgium

Ronald BLAB
Institute of Transportation
Vienna University of Technology
Austria

Thomas BLES
Deltares
Delft
The Netherlands

Adrian BOHANE
TRE Canada Inc.
Vancouver
Canada

Benjamin BRINGIER
Laboratoire XLIM-SIC, UMR CNRS
University of Poitiers
France

Brian BRUCKNO
Virginia Department of Transportation
Richmond
USA

Nicolas BUECHE
Traffic Facilities Laboratory
Ecole Polytechnique Fédérale de
Lausanne
Switzerland

Elizabeth CAMPBELL
Virginia Department of Transportation
Richmond
USA

Dermot B. CASEY
Atkins
Nottingham
UK

Josep Maria CARBONELL
International Center for Numerical
Methods in Engineering (CIMNE)
Barcelona
Spain

Sean CASSIDY
Lagan Asphalt
Dublin
Ireland

Gary CHAI
Griffith School of Engineering
Griffith University
Australia

Sylvain CHATAIGNER
French Institute of Science and Technology for Transport, Development and Networks (IFSTTAR)
Bouguenais
France

Barbara CHIMANI
Central Institute for Meteorology and Geodynamics (ZAMG)
Vienna
Austria

Sanaul CHOWDHURY
Griffith School of Engineering
Griffith University
Australia

Paul CROMPTON
ARTTIC
Brussels
Belgium

Mauro D'APUZZO
University of Cassino and Southern Lazio
Cassino
Italy

Isabelle DE KEYZER
UIC
Paris
France

Edgar DE LEÓN IZEPPI
Center for Sustainable Transportation Infrastructure
VTTI
Blacksburg, Virginia
USA

Frédéric DELFOSSE
Eurovia Research Centre
Mérignac
France

François DEPARDON
LCPC Experts
Marne-la-Vallée
France

Klaus DIEPLINGER
Via Donau
Vienna
Austria

Bjorn Kristoffer DOLVA
Norwegian Public Roads Administration
Oslo
Norway

Ivan DROUADAINE
Eurovia Research Centre
Mérignac
France

André-Gilles DUMONT
Traffic Facilities Laboratory
Ecole Polytechnique Fédérale de Lausanne
Switzerland

Jean-Paul DURANTHON
CEREMA
DTCE
Bron
France

Bernard ECKMANN
Technical Department
EUROVIA
Rueil-Malmaison
France

Mahmoud ELNASRI
Nottingham Transportation Engineering Centre (NTEC)
Department of Civil Engineering
University of Nottingham
UK

Michael ENGELS
Wirtgen GmbH
Germany

Azzurra EVANGELISTI
University of Cassino and Southern Lazio
Cassino
Italy

Philippe ÉVENAT
ESTAR
La Richardais
France

Giacomo FALORNI
TRE Canada Inc.
Vancouver
Canada

Stéphane FAUCON DUMONT
Eurovia Research Centre
Mérignac
France

Gerardo W. FLINTSCH
Center for Sustainable Transportation Infrastructure
VTTI
Blacksburg, Virginia
USA

Zuzana FORMANOVÁ
Czech Technical University
Prague
Czech Republic

Tobias FUCHS
German National Meteorological Service (DWD)
Offenbach
Germany

Kenneth GAVIN
Centre for Critical Infrastructure Research
University College Dublin
Ireland

Amanda GIBNEY
University College Dublin
Ireland

Konstantinos GIANNAKOS
Civil Engineering Dept.
University of Thessaly
Volos
Greece

Kevin GILBERT
Shell Bitumen
Manchester
UK

António GOMES CORREIA
Department of Civil Engineering
University of Minho
Guimarães
Portugal

José Manuel GONZALEZ
International Center for Numerical Methods in Engineering (CIMNE)
Barcelona
Spain

Luc GOUBERT
Belgian Road Research Centre (BRRC)
Brussels
Belgium

James R. GRENFELL
Nottingham Transportation Engineering Centre (NTEC)
University of Nottingham
UK

Hong GUAN
Griffith School of Engineering
Griffith University
Australia

Alexander HABERL
Institute of Transportation
Vienna University of Technology
Austria

Pierre HANKACH
French Institute of Science and Technology for Transport, Development and Networks (IFSTTAR)
Bouguenais
France

Przemysław HARASIM
Road and Bridge Research Institute (IBDiM)
Warsaw
Poland

Thomas HARTL
Via Donau
Vienna
Austria

Andreas HARTMANN
University of Twente
Enschede
The Netherlands

Katrin HASELBAUER
Institute of Transportation
Vienna University of Technology
Austria

Rayya HASSAN
Swinburne University of Technology
Melbourne
Australia

Carina HERRMANN
Bundesanstalt für Straßenwesen
Bergisch Gladbach
Germany

Inge HOFF
Norwegian University of Science and Technology
Trondheim
Norway

Markus HOFFMANN
Institute of Transportation
Vienna University of Technology
Austria

Edward HOPPE
Virginia Transportation Research Council
Charlottesville
USA

Pierre HORNYCH
French Institute of Science and Technology for Transport, Development and Networks (IFSTTAR)
Bouguenais
France

Jan HRADIL
CTU FCE
Department of road structures
Prague
Czech Republic

Nicolas HUYBRECHTS
CETMEF
Margny-lès-Compiègne
France

Aleksander IPAVEC
Slovenian National Building and Civil Engineering Institute (ZAG)
Ljubljana
Slovenia

Bernard JACOB
French Institute of Science and Technology for Transport, Development and Networks (IFSTTAR)
Marne-la-Vallée
France

Patrick JOFFRIN
French Institute of Science and Technology for Transport, Development and Networks (IFSTTAR)
Bron
France

Björn KALMAN
Swedish National Road and Transport Research Institute (VTI)
Linköping
Sweden

Carsten KARCHER
Karlsruhe Institute of Technology
Germany

Samer W. KATICHA
Center for Sustainable Transportation Infrastructure
VTTI
Blacksburg, Virginia
USA

Greg KELLY
Griffith School of Engineering
Griffith University
Australia

Jean-Pierre KERZRÉHO
French Institute of Science and Technology for Transport, Development and Networks (IFSTTAR)
Bouguenais
France

Majdi KHOUDEIR
Laboratoire XLIM-SIC, UMR CNRS
University of Poitiers
France

Darko KOKOT
Slovenian National Building and Civil Engineering Institute (ZAG)
Ljubljana
Slovenia

Jozef KOMAČKA
University of Žilnia
Slovakia

Jean-Bernard KOVARIK
Ministry of Infrastructure, Transport and the Sea
Paris La Défense
France

Kathrin KUBANEK
Karlsruhe Institute of Technology
Karlsruhe
Germany

Daniel A. KUCHMA
University of Illinois at Urbana-Champaign
Illinois
USA

Kyle KWIATKOWSKI
University of Colorado
USA

Anne LALAGÜE
SINTEF Building and Infrastructure Trondheim
Norway

Sabine LARGEAUD
Eurovia Research Centre
Mérignac
France

Francesca LA TORRE
University of Florence
Italy

Matthew A. LEBENS
Minnesota Department of Transportation Maplewood
Minnesota
USA

Greet LEEGWATER
TNO
Delft
The Netherlands

François LEMAÎTRE
ONERA
Toulouse
France

Stanislav LENART
ZAG Ljubljana
Slovenia

Philippe LEPERT
French Institute of Science and Technology for Transport, Development and Networks (IFSTTAR)
Bouguenais
France

Philippe LE STER
ESTAR
La Richardais
France

Fabien LÉTOURNEAUX
SYSTRA
Paris
France

Oliver LIN
VicRoads
Melbourne
Australia

Johan LITZKA
Technical University of Vienna
Austria

Duo LIU
Road, Railway and Airfield Construction
Technical University of Munich
Germany

Gang LIU
Delft University of Technology
The Netherlands

Jean-Pierre MAGNAN
French Institute of Science and Technology for Transport, Development and Networks (IFSTTAR)
Marne-la-Vallée
France

Dominika MALISZEWSKA
Road and Bridge Research Institute (IBDiM)
Warsaw
Poland

Maciej MALISZEWSKI
Road and Bridge Research Institute (IBDiM)
Warsaw
Poland

Christoph MATULLA
Central Institute for Meteorology and Geodynamics (ZAMG)
Vienna
Austria

Ciaran MCNALLY
University College Dublin
Ireland

Goran MLADENOVIC
Faculty of Civil Engineering
University of Belgrade
Serbia

Konrad MOLLENHAUER
University of Kassel
Germany

Leto MOMM
Federal University of Santa Catarina
Florianópolis
Brazil

Jessica MORGAN
TRE Canada Inc.
Vancouver
Canada

Joachim NAMYSLO
German National Meteorological Service (DWD)
Offenbach
Germany

Anh-Duc NGUYEN
Technische Universitaet Darmstadt
Germany

Mai-Lan NGUYEN
French Institute of Science and Technology for Transport, Development and Networks (IFSTTAR)
Bouguenais
France

Cliff NICHOLLS
Transport Research Laboratory
Wokingham
UK

Vittorio NICOLOSI
University of Rome Tor Vergata
Rome
Italy

Erik NIELSEN
Danish Road Directorate
Hedehusene
Denmark

Ayman M. OKEIL
Louisiana State University
Baton Rouge, Louisiana
Louisiana
USA

Gustavo Garcia OTTO
Federal University of Santa Catarina
Florianópolis
Brazil

Abdellatif OUAHSINE
University of Technology of Compiègne
France

Jean-Marc PAUL
French Institute of Science and Technology for Transport, Development and Networks (IFSTTAR)
Bouguenais
France

Manuel PARENTE
Department of Civil Engineering
University of Minho
Guimarães
Portugal

Dirk PEREBOOM
Deltares
Delft
The Netherlands

Sergio PEREZ
Bombardier Transportation
Mannheim
Germany

Philippe POILANE
Technical Department
Eurovia Centre Ouest
Nantes
France

Luke J. PRENDERGAST
Centre for Critical Infrastructure Research
University College Dublin
Ireland

Marc QUIERTANT
French Institute of Science and Technology for Transport, Development and Networks (IFSTTAR)
Marne-la-Vallée
France

Eva REMIŠOVA
University of Žilnia
Slovakia

Sami RIZKALLA
Department of Civil, Construction and Environmental Engineering
North Carolina State University
Raleigh, North Carolina
USA

Arnaud ROLLAND
CEREMA
Direction Territoriale Nord-Picardie
Haubourdin
France

Franziska SCHMIDT
French Institute of Science and Technology for Transport, Development and Networks (IFSTTAR)
Marne-la-Vallée
France

Laurent SCHMITT
UIC
Paris
France

Philippe SERGENT
CETMEF
Margny-lès-Compiègne
France

Ji SHENGCHENG
University of Technology of Compiègne
France

Reza SHAHMOHAMMADI
Lagan Asphalt
Dublin
Ireland

Markus SIMONER
Via Donau
Vienna
Austria

Petr SLABÝ
CTU FCE
Department of road structures
Prague
Czech Republic

Hassan SMAOUI
University of Technology of Compiègne
France

Irina STIPANOVIĆ OSLAKOVIĆ
University of Twente
The Netherlands

Michael STUECHELI
University of Virginia
Charlottesville
USA

Jan SUDA
Czech Technical University
Prague
Czech Republic

Amir TABAKOVIĆ
University College Dublin
Ireland

Bruno TAILLIS
Technical Department
Eurovia Sud-Ouest
Mérignac
France

Han TER MAAT
Wageningen University
The Netherlands

Vikas THAKUR
Norwegian Public Roads Administration
Oslo
Norway

Amutha THANANJEYAN
VicRoads
Melbourne
Australia

Nick THOM
Nottingham Transportation Engineering Centre (NTEC)
Department of Civil Engineering
University of Nottingham
UK

Christiaan TOLLENAAR
M+P Consulting Engineers
Vught
The Netherlands

Jean-Michel TORRENTI
French Institute of Science and Technology for Transport, Development and Networks (IFSTTAR)
Marne-la-Vallée
France

François TOUTLEMONDE
French Institute of Science and Technology for Transport, Development and Networks (IFSTTAR)
Marne-la-Vallée
France

Marjan TUŠAR
National Institute of Chemistry
Ljubljana
Slovenia

Michal UHLÍK
CTU FCE
Department of Road Structures
Prague
Czech Republic

Andrea VACCARI
University of Virginia
Charlottesville
USA

Marko VAJDIĆ
Institut IGH d.d.
Zagreb
Croatia

Jan VALENTIN
Czech Technical University
Prague
Czech Republic

Amir Mattar VALENTE
Federal University of Santa Catarina
Florianópolis
Brazil

Wouter VAN BIJSTERVELD
Geotecnia y cimientos (GEOCISA)
Madrid
Spain

Gijsjan VAN BLOKLAND
M+P Consulting Engineers
Vught
The Netherlands

Ronald VAN LOON
M+P Consulting Engineers
Vught
The Netherlands

Kees VAN MUISWINKEL
Rijkswaterstaat Water, Traffic and Environment
Delft
The Netherlands

Rudi VAN STADEN
College of Engineering and Science
Victoria University
Melbourne
Australia

Aikaterini VARVERI
Delft University of Technology
The Netherlands

Axel WALTHER
Institut für Straßenwesen
Technische Universität
Braunschweig
Germany

Matthew WAYMAN
Transport Research Laboratory
Wokingham
UK

Alfred WENINGER-VYCUDIL
PMS-Consult GmbH
Engineering Office for
Traffic and Infrastructure
Vienna
Austria

Michael WISTUBA
Institut für Straßenwesen
Technische Universität
Braunschweig
Germany

Xiaoyi ZHOU
French Institute of Science and Technology for Transport, Development and Networks (IFSTTAR)
Marne-la-Vallée
France

Adam ZOFKA
Road and Bridge Research Institute (IBDiM)
Warsaw
Poland

Index

A, C, D

E, F

G, H, I, L

P, R,

S, T

V, W

Contents for Volume 5A

Printed and bound by CPI Group (UK) Ltd, Croydon, CR0 4YY

15/09/2026

14962356-0002